Essentials of Semiconductor Device Physics

Essentials of Semiconductor Device Physics

EMILIANO R. MARTINS

Department of Electrical and Computer Engineering
University of São Paulo, Brazil

Registered Offices
John Wiley & Sons, Inc., 111 River Street, Hoboken, NJ 07030, USA
John Wiley & Sons Ltd, The Atrium, Southern Gate, Chichester, West Sussex, PO19 8SQ, UK

Editorial Office
The Atrium, Southern Gate, Chichester, West Sussex, PO19 8SQ, UK

For details of our global editorial offices, customer services, and more information about Wiley products visit us at www.wiley.com.

Wiley also publishes its books in a variety of electronic formats and by print-on-demand. Some content that appears in standard print versions of this book may not be available in other formats.

Library of Congress Cataloging-in-Publication Data

Name: Martins, Emiliano R., author. | John Wiley & Sons, publisher.
Title: Essentials of semiconductor device physics / Emiliano R. Martins.
Description: Hoboken, NJ : Wiley, 2022. | Includes bibliographical
   references and index.
Identifiers: LCCN 2021059908 (print) | LCCN 2021059909 (ebook) | ISBN
   9781119884118 (paperback) | ISBN 9781119884125 (adobe pdf) | ISBN
   9781119884132 (epub)
Subjects: LCSH: Semiconductors.
Classification: LCC TK7871.85 .M345 2022 (print) | LCC TK7871.85 (ebook)
   | DDC 621.3815/2–dc23/eng/20220112
LC record available at https://lccn.loc.gov/2021059908
LC ebook record available at https://lccn.loc.gov/2021059909

Cover Design: Wiley
Cover Image: © 7activestudio/Getty Images

Set in 11/13pt Computer Modern by Straive, Pondicherry, India
Printed and bound by CPI Group (UK) Ltd, Croydon, CR0 4YY

C9781119884118_200622

Contents

Preface

This book presents the essential physical processes underlying the operation of semiconductor devices, with an emphasis on the physics of **p-n** junctions.

The book is written for undergraduate students taking a short introductory course in semiconductor devices, as typically required in electrical engineering, physics, and material sciences.

A major challenge in teaching semiconductor physics to undergraduate students is that they often do not have an appropriate background in statistical physics. It is not rare for students to embark on a course in semiconductors with no previous conception, for example, of what a Fermi level is. Most books on semiconductors, however, assume a previous background in statistical physics, or even quantum mechanics, thus making them inaccessible to many undergraduates.

This book is an attempt to solve this problem. Here, I assume that the reader has no previous knowledge of statistical physics or quantum mechanics. The main concepts of statistical physics required for a mature understanding of semiconductor device physics are thus provided in Chapter 1. Some very basic concepts of quantum mechanics are also introduced along the way, when necessary. The content of the first chapter is one of the key features that sets this book apart from standard texts in semiconductor physics.

The book opens with a chapter on statistical physics, where the concepts of entropy, temperature, and chemical potential are introduced. Special care is taken to introduce the chemical potential and the Fermi level in an intelligible way. Overall, the concepts are treated rigorously, though I try not to compromise clarity with excessive rigor. The Fermi–Dirac distribution is deduced by means of an example, which I find to be much more intuitive than the traditional derivation based on combinatorial analysis. Finally, the chapter on statistical physics closes with a short introduction to the transport processes relevant to the operation of **p-n** junctions.

The physics of semiconductors per-se are introduced in Chapter 2, relying heavily on the concepts of Chapter 1. The essentials of band theory and hole transport are discussed, followed by calculations of charge carrier concentrations. To emphasise the core ideas behind these calculations, the latter are performed twice: first using a brute force method, and then using the density of states. The chapter closes with a description of doping and its effect on the Fermi level.

Chapter 3 introduces the **p-n** junction, which is the main goal of the text. The emphasis is on the role of the Fermi level, and on describing how the relationship between chemical and electrostatic potentials define the electrical properties of a **p-n** junction. The electric field in the depletion region is calculated assuming uniform doping and the chapter closes with a derivation of the current vs voltage relationship in a **p-n** junction (Shockley equation). As this is an introductory text, non-idealities are only mentioned, but not described in detail.

Chapters 4 and 5 show two important applications of **p-n** junctions.

Chapter 4 introduces photovoltaic devices, with emphasis on solar cells. The physics of solar cells are discussed, and the current vs voltage relationship is explained by means of an equivalent circuit. Finally, the chapter closes with a brief exposition on the efficiencies of solar cells and the physical origins of their limitations.

Chapter 5 introduces the two main types of transistor: Bipolar Junction Transistors (BJTs) and Metal Oxide Semiconductor Field Effect Transistors (MOSFETs). The basic physical principles governing the operation of both types of transistor are presented, and three paradigmatic applications of transistors are discussed: the transistor as an amplifier, the transistor as an electronic switch, and their applications in logic gates.

The text assumes that the reader has no previous knowledge of statistical physics or quantum mechanics, but it assumes a basic knowledge of electrostatics.

About the author

Emiliano earned his PhD in Physics from the University of St. Andrews (UK) in 2014 and has been teaching semiconductors in the Department of Electrical and Computer Engineering at the University of São Paulo (Brazil) since 2016. He is passionate about learning and teaching, about books, coffee and British comedy. He also finds it a bit weird to talk about himself in the third person.

Acknowledgments

The raison d'être of all my endeavours lies within three persons: my wife Andrea, and my two children Tom and Manu. Without you, no book would be written, no passion would be meaningful.

I am grateful to my parents Elcio and Ana for teaching me what really matters, and to my sister Barbara for showing me the easiest way to learn what really matters.

I also thank all my students who contributed with invaluable feedback over all these years. You were the provers guiding the recipe of the pudding. If I have lived up to your expectations, then I have done my job well.

I am grateful to the Wiley team who brought the manuscript to life: Skyler Van Valkenburgh, Sakthivel Kandaswamy, Richard Walshe and Martin Preuss.

Last, but certainly not least, I thank my editor Jenny Cossham for believing in this project.

About the companion website

This book is accompanied by a companion website:

www.wiley.com/go/martins/essentialsofsemiconductordevicephysics

This website includes:

- Solutions
- PowerPoint Slides

1

Concepts of statistical physics

Learning objectives

*The two most important concepts of statistical physics pertaining to the physics of semiconductor devices are the electrochemical potential, also known as the Fermi level, and the Fermi–Dirac distribution. As soon as the electrochemical potential is introduced, I will show you a prototype of a **p-n** junction, and emphasize the key role played by the electrochemical potential in establishing the equilibrium properties of **p-n** junctions. The Fermi–Dirac distribution is another key concept treated in this chapter. It will be derived in a way that highlights its assumptions and domains of applicability, and it will be used in the next chapter to find the concentration of charges in semiconductors. This chapter also introduces two types of currents relevant to **p-n** junctions: the drift current and the diffusion current.*

1.1 Introduction

We begin our introduction to statistical physics with the following principle: *the state property of a macroscopic system in equilibrium is always the most probable.* Our first job is to make sense of this principle.

First, we need to understand the meaning of the key terms: system, state and equilibrium. We shall begin with the simplest term: system.

A system is just the group of stuff we are interested in, including the stuff that we are not directly interested in but that affects the stuff we are interested in. For example, if we are interested in studying the interaction

Essentials of Semiconductor Device Physics, First Edition. Emiliano R. Martins.
© 2022 John Wiley & Sons Ltd. Published 2022 by John Wiley & Sons Ltd.
Companion website: www.wiley.com/go/martins/essentialsofsemiconductordevicephysics

between two gases, then the two gases will be part of our system. We might also need to consider the recipient where the gases are contained as part of our system. Furthermore, if the gases are not isolated from the external environment, then we also need to consider it as part of our system. So, "system" is just the technical term for the stuff we are interested in.

The idea of states, however, is more subtle. In physics, to specify the state of a system is to give the information one needs to predict the evolution of the system. As such, one can understand the state as being the initial condition of the system. In Newtonian physics for example, the state is given by the positions and momenta of all the particles in the system.

A state has a set of well-defined properties. For example, if we specify the positions and momenta of all the particles, then we can know the energy, because from the positions we can work out the potential energy, and from the momenta we can work out the kinetic energy. So, the energy is a property of the state. Notice that more than one state can have the same property (for example, two different states can have the same energy).

There is, however, an important difference between classical and quantum states: in quantum mechanics, one often finds a discrete number of "allowed states", whereas in classical physics the states form a continuum.

Figure 1.1 A classical system comprising a ball of mass m oscillating on a frictionless ramp may acquire a continuum of states and energies.

This difference can be illustrated using a simple example. Consider a classical system comprising a ball of mass m in a U-shaped potential (a skate ramp), as illustrated in Figure 1.1. When the ball is dropped from a height h, it then accesses a continuum of states, i.e., the position and momenta are changed continually as the ball oscillates. Furthermore, there is no restriction on the energy of the ball, as the ball can be dropped, in principle, from any height.

In quantum physics, it often happens that only a discrete number of states, and hence of energies, can be accessed by the system. It is as though the ball could only sit on discrete shelves of specific heights, as loosely illustrated in Figure 1.2.

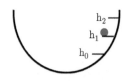

Figure 1.2 In a quantum system, it may happen that only discrete states are available, as if the ball could only sit at discrete heights, acquiring discrete levels of energy.

It is not difficult to understand the origin of this restriction on the available states of quantum systems. As a matter of fact, this phenomenon is not restricted to quantum physics; it is, in fact, a wave phenomenon, which arises when a wave is confined. An optical fibre, for example, is a system that confines electromagnetic waves; so, only a discrete set

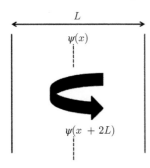

L

$\psi(x)$

$\psi(x + 2L)$

Figure 1.3 A confined wave can only exist if the phase accumulated in a round trip is a multiple of 2π.

of states is allowed in an optical fibre. There are two types of optical fibres: monomode and multimode. Only one state of the electromagnetic field is allowed in a monomode fibre, whereas a multimode fibre admits more than one state, but still only a discrete number of states.

To illustrate the origin of the discretization, consider a one-dimensional cavity (i.e., a box) where a wave is confined, as shown in Figure 1.3. The wave can be of any nature, but to make the example more realistic, let us suppose that the cavity is formed by parallel mirrors and that the wave is electromagnetic.

When the wave propagates a full round trip, it returns upon itself. Therefore, to exist inside the cavity, it must accumulate a round-trip phase equal to a multiple of 2π, otherwise it would "interfere destructively with itself".

This restriction can easily be shown mathematically by considering a plane wave ψ, whose spatial dependence $\psi(x)$ is described as:

$$\psi(x) = exp\,(ikx) \tag{1.1}$$

where i is the imaginary number and k is the wave propagation constant (or wavenumber). Notice that the temporal dependence was omitted, as we are interested only in the phase accumulation due to spatial propagation.

Next, we need to determine the phase accumulated in a round trip. First, notice that the wave is reflected by a mirror twice. In each reflection, it picks up a π phase shift, adding up to a total phase due to reflection of 2π. Therefore, the two mirrors do not contribute to a phase difference. In addition to the phase picked up upon reflection, the wave accumulates a phase due to propagation. If the cavity length is L, then the wave propagates a distance of $2L$ in a round trip; thus:

$$\psi(x + 2L) = exp\,[ik(x + 2L)] = \psi(x)\,exp\,(ik2L) = \psi(x)\,exp\,(i\Delta\varphi) \tag{1.2}$$

As is apparent in Equation (1.2), the phase difference $\Delta\varphi$ between the wave at the point x and the wave at the point $x + 2L$ is $\Delta\varphi = k2L$. But since the wave returns upon itself after a round-trip propagation, the positions x and $x + 2L$ are, actually, the same position, i.e.:

$$\psi(x) = \psi(x + 2L)$$

The relation above is sometimes called the "self-consistency relation". As is evident in Equation (1.2), the self-consistency relation can only be satisfied if $\Delta\varphi$ is a multiple of 2π. In other words, the cavity imposes a restriction on the propagation constant, which must satisfy the condition:

$$k2L = c2\pi \tag{1.3}$$

where c is a natural number.

In the language of quantum mechanics, Equation (1.3) is affirming that the "allowed states" of the systems are those whose propagation constants are multiples of π/L, that is:

$$k = c\frac{\pi}{L} \tag{1.4}$$

The first state is given by $c = 1$, the second state is given by $c = 2$, and so on. Notice that we have an infinite number of states (as c can be any natural number), but that they are discrete (by the way, negative integers do not count because they describe the same state of their positive counterpart).

The imposition of the self-consistency relation by a cavity is a general property of wave systems (see Box 1). As it turns out, quantum mechanics describes particles as waves (the famous particle wavefunctions), so these constraints on the propagation constants also apply to particles trapped in a cavity. The most common example is that of an electron trapped in a box. The fancy term for a box used to trap electrons is "quantum well". So, a quantum well is a cavity in quantum mechanics.

Now we focus attention on the consequences of the imposition of the self-consistency relation on an electron trapped in a 1D quantum well. For that, suppose that Equation (1.1) represents an electron wavefunction, and that this electron is trapped in a 1D quantum well. The quantum well imposes the self-consistency relation, which means that Equation (1.4) is also valid for the electron's propagation constant. However, according to the famous de Broglie's equation, the electron's mechanical momentum (call it p) is proportional to the propagation constant:

$$p = \frac{h}{2\pi}k \tag{1.5}$$

where h is the even more famous Planck's constant (if you are wondering where Equation (1.5) came from, it is kind of embedded in the postulates of quantum mechanics). Expressing the self-consistency relation (Equation (1.4)) in terms of the mechanical momentum, we get:

$$p = c\frac{h}{2L} \tag{1.6}$$

This equation expresses the fact that the mechanical momentum of an electron trapped in a quantum well is restricted to a discrete set of values.

Box 1   Guitar strings as another example of wave confinement

As another example of discretization arising from wave confinement, consider a system consisting of a string attached on two sides (just like in a guitar), as shown in Figure B1.1.

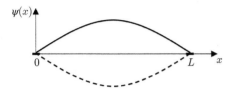

Figure B1.1 Example of a confined wave system: a string attached on both sides, just like in a guitar.

The distance between the two points of attachment is the length of the cavity, and the height of the string is the amplitude of the wave, which again I denote by $\psi(x)$ (recall that we don't need the time dependence to understand this effect, so I ignore it).

This system supports standing waves, whose spatial dependence is of the form:

$$\psi(x) = A sin(kx) + B cos(kx)$$

The parameters A and B depend on the initial conditions. The confinement of the wave, that is, the attachment at points $x = 0$ and $x = L$, imposes two boundary conditions:

$$\psi(0) = 0$$

and

$$\psi(L) = 0$$

Applying the first boundary condition, we find

$$\psi(0) = A sin(k0) + B cos(k0) = B = 0$$

Thus:

$$\psi(x) = A sin(kx)$$

Now, applying the second boundary condition, we find:

$$\psi(L) = A sin(kL) = 0$$

One possible solution is, of course, $A = 0$, but that is the trivial solution describing the string at rest (nobody playing the guitar). The interesting solutions are those for $A \neq 0$, which obviously requires that

$$sin(kL) = 0$$

(Continued)

(Continued)

Recall that the sine function is zero when its argument is a multiple of π. Thus, the condition $sin(kL) = 0$ entails the condition:

$$kL = c\pi$$

where c is a natural number.

Thus, we conclude that the possible values of k satisfy:

$$k = c\frac{\pi}{L}$$

This is the same requirement we obtained for the system "electromagnetic waves trapped between two parallel mirrors".

We can express this restriction in terms of the kinetic energy E by using the well known relationship between kinetic energy and momentum in classical mechanics:

$$E = \frac{p^2}{2m} \tag{1.7}$$

where m is the electron mass.

Combining Equation (1.7) and Equation (1.6), we find that the allowed energies of an electron trapped in a quantum well are given by:

$$E_c = \frac{h^2 c^2}{8L^2 m} \tag{1.8}$$

where the subindex c was inserted to differentiate between energies given by different values of c.

Notwithstanding its simplicity, Equation (1.8) is very useful and we will often refer to it in this book. Each value of c denotes a different energy level, so Equation (1.8) describes the allowed energy levels of the system. One important feature captured by Equation (1.8) is that, if the quantum well is made sufficiently small, then the energy difference between adjacent energy levels becomes comparable to the energy of photons (photons are the quanta of light). Consequently, energy conservation allows electrons to absorb or emit photons when changing states (this transition between states is also called a "quantum jump"). Nowadays, there are many techniques to produce cavities designed to control the emission and absorption of light. These systems are commonly called "quantum dots", and they find important applications in many modern technologies, ranging from cancer treatment to displays. The screens based on Quantum dot Light Emitting Diodes (QLED) are a flashy example of application of quantum dots.

It is important to emphasize that each value of c denotes a different allowed state, and that E_c in Equation (1.8) is the energy (hence the property) of state c. Furthermore, notice that, in this example, two different states necessarily have two different values of energy. This is not generally true, however. Indeed, as we will see shortly, in a 3D cavity, more than one state can have the same energy.

These considerations suffice to illustrate the concepts of state and state properties. If you go back to the discussion in the beginning of this section, you will see that we still need to establish the notion of equilibrium. But now that we know what a system is, what a state is and what a state property is, it gets easy to define equilibrium: we say that a system is in equilibrium when its properties are no longer altered. For now, the only state property we are interested in is the energy. Thus, in the example of the two interacting gases, the properties (the energies) of each gas change as the gases interact. When they reach equilibrium, however, the properties stop changing. Notice that the gases continue to interact after reaching equilibrium, but now in a such a way that the properties are constant in time.

But why? What is so special about these properties in equilibrium that allows them to stop changing? The answer to these questions is in the probabilistic principle enunciated at the beginning of this section: *the state property of a macroscopic system in equilibrium is always the most probable.*

Let us take some time to reflect upon this principle. It is essentially affirming that the property of the system in equilibrium is the most probable. Notice that the principle could be recast as: *a macroscopic system whose property is no longer altered in time has the most probable property.* This second form of the principle is far less elegant, but it is useful in highlighting the fact that the term equilibrium only conveys the fact that the property does not change anymore.

So, what exactly is this principle affirming? Remember that a state is the physical condition of a system. If the system is a gas, its state is determined by the positions and momenta of all molecules of the gas. Each combination of position and momentum gives a different state. But each state has a well-defined property, that is, a well-defined energy (later we will consider another property, but for now just the energy suffices). In this case, the principle is stating that, as the two gases interact, their energies will keep changing until they reach their most probable value. After that, the energies do not change anymore. More precisely, still considering the example of the two interacting gases, and isolated from the universe, suppose that, in the beginning of the interaction, the energy of the first gas is $U1_0$, while the energy of the second gas is $U2_0$. The gases exchange energy as they interact, but since energy is conserved, they always satisfy $U1 + U2 = U1_0 + U2_0$, where $U1$ and $U2$ are the energies of gas 1 and 2 at any moment. Therefore, the accessible states of the system are those whose properties satisfy $U1 + U2 = U1_0 + U2_0$. According to our principle, once equilibrium is reached, then $U1$ and $U2$ are fixed: they do not change in time anymore. If you measure the energy of gas 1 after equilibrium, you will always get the value $U1$ (and $U2$ for gas 2). Furthermore, these values, $U1$ and $U2$, after equilibrium, are the most probable (notice that, due to energy conservation, $U1$ and $U2$ are not independent of each other). So, what the principle is essentially affirming is "take all the values of energies that satisfy energy conservation; count how many states have each of these values of energy as a property; the value of energy with the largest number of states is the value of equilibrium (because the value of energy that has the largest number of states is the most probable). So, after some time, the gases 1 and 2 will have these most probable energies, and after that the energies do not change any more".

These considerations lead to three questions: how can we find the energies and the possible states; what does the term "most probable property" really mean; and why does the system evolve to these most probable properties and stays there? These three questions are better addressed by means of an example.

For the sake of simplicity, let us use as an example a binary system, that is, a system whose elements have only two states. We can pretend that these elements are coins, and that *heads up* denotes one state, and *tails up* denotes the other. To make the example more realistic, we need to ascribe a property to these states. Let us say that a coin with heads up has energy $U = 0$, and a coin with tails up has energy $U = 1$. Let us also define N as the number of elements (coins) in our system.

So, say that we have a system of two coins ($N = 2$). What are the possible states of the system? The system is composed of elements (the coins), and the elements have two states each (*heads* or *tails*). So, the states of the *system* are composed of combination of states of the elements (try to avoid confusing the state of the system with the state of the elements). In our example, denoting heads by H and tails by T, the *system* has four different states: HH, HT, TH, TT. And, according to our scheme, the energies of these states are, respectively, 0, 1, 1, 2. Thus, there are three possible energies, $U = 0$, $U = 1$, and $U = 2$. And notice that, because two states have energy $U = 1$, then $U = 1$ is the most probable energy in this example.

In our example, the differences between the probabilities are not great. Indeed, in a coin tossing experiment involving two coins, the probability of getting an energy $U = 0$ is 25%, of getting an energy $U = 1$ is 50% and of getting an energy $U = 2$ is 25%. So, in our example, the most probable energy is not tremendously more probable than the other energies. The probabilities are similar in our example because it deals with a small, or microscopic, system. Indeed, a system with only two elements is quite small.

Now we shall see that, as the number of elements increases, then the most probable result (the energy, or, equivalently, the number of tails up) becomes overwhelmingly more probable than all the other possible results.

To see that this is indeed the case, let us keep our example of coins, but let us check how the probabilities of the energies (tails up) evolve as the number of coins increases. If there are N coins in the system, then there are 2^N possible states. We define the number of states with M tails up in a system of N coins as $A(N, M)$. So, $A(N, M)$ is the number of ways that N coins can be arranged with M tails up. From basic combinatorial analysis, we know that:

$$A(N, M) = \frac{N!}{M!(N - M)!} \tag{1.9}$$

So, the probability $P(N, M)$ of finding M tails up in a coin toss involving N coins is:

$$P(N, M) = \frac{A(M, N)}{2^N} = \frac{N!}{2^N M!(N - M)!} \tag{1.10}$$

We can use either Equation (1.9) or Equation (1.10) to study the evolution of the probability of finding M coins with tails up in a coin toss experiment involving N coins. Thus, three histograms are shown in Figure 1.4 for systems with an increasingly larger number of coins. Notice that, as the system's size increases (N increases), the histogram gets more concentrated around its most probable value (and the most probable value is $M = N/2$, that is, half of the coins with tails up). This means that, for large systems, the probability of finding the most probable value is overwhelmingly higher than the probability of finding any other value. For example, the histogram of the system with $N = 10^6$ coins already resembles a delta of Dirac function.

It is instructive to quantify the narrowing of the probability distribution as a function of the system's size. For that, one can define the width ΔM of the histogram as $\Delta M = M_{+e} - M_{-e}$, where M_{+e} and M_{-e} are, respectively, the value of M above and below the most probable value at which the probability falls by $1/e$ of its maximum value. Thus, ΔM is a measure of the spread around the most probable value within which the histogram retains a significant probability. For large systems, from Equation (1.9) and with the help of the Stirling approximation, the following approximation can be derived:

$$\frac{\Delta M}{M_{mp}} \approx 4\sqrt{\frac{1}{2N}} \tag{1.11}$$

where M_{mp} is the most probable value (that is, $M_{mp} = N/2$). Thus, the left-hand side of Equation (1.11), called the "fractional width" of the histogram,

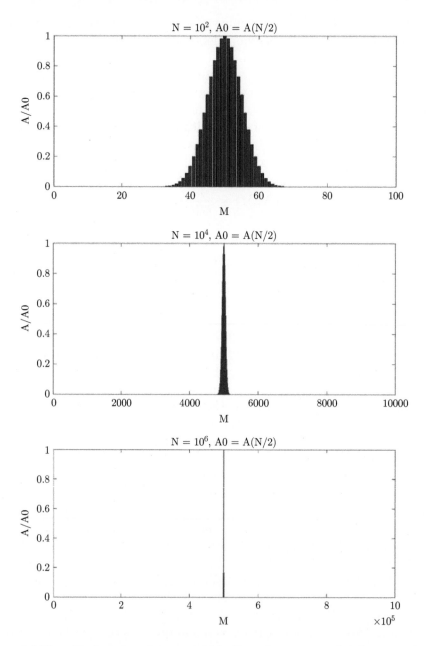

Figure 1.4 Normalized number of states with M tails up for systems with different number of coins N. The histograms are normalized to its maximum value A0.

is a measure of the relative spread around the most probable value for which there is a meaningful probability.

As expected, according to Equation (1.11), the larger the system is (that is, the larger N is), the smaller its fractional width is. For example, for $N = 10^6$, the fractional width is of the order of 10^{-3}. This means that, in a coin toss with $N = 10^6$ coins, one can claim that the number of coins with tails up will be $M_{mp} \pm 10^{-3} \times M_{mp}$. This is an error of only 0.1%. Thus, it is pretty safe to affirm that, for $N = 10^6$, half of the coins will be landing tails up.

Speaking more informally, our example is telling us that, in a coin toss experiment involving 10^6 coins, one will not be surprised to find, say, 500,005 coins with tails up, since 500,005 is a number very close (in percentual terms) to the most probable number (which is 500,000 coins with tails up). But, if in a coin toss experiment involving 10^6 coins, one finds 250,000 coins landing tails up, one will be justified in suspecting one has been tricked. Indeed, since the fractional width in this example is 10^{-3}, the expected variation is between $500,000 - 500,000 \times 10^{-3} = 499,500$ and $500,000 + 500,000 \times 10^{-3} = 500,500$ tails up. Anything outside this range is too improbable.

Therefore, if one day your health and safety depend on you having to guess the number of tails up in an experiment involving 10^6 coins, a guess of 500,000 will most likely save your skin. And that is true because the most probable property of the system is overwhelmingly more probable than the other properties (remember that, in our example, the property is the number of tails up, which represents the energy, and the most probable property is half of the tails up, that is $U = N/2$). And this is true for the simple reason that there are many more states (remember that each arrangement of coins is a state) with the most probable property than with the other properties (that is, there are many more states with half of the coins with tails up, than with other numbers of tails up).

Now, imagine that you arrive in a room with one thousand coins spread out on the floor. If it looks like that half of them have tails up, and half tails down, then you are justified in thinking that they were spread randomly. If, on the other hand, you find all the coins with tails up, then you will guess that someone purposefully organized them. This intuition is grounded on the fact that you know that there are many more states for the "disorganized" property than for the "organized" property, so the "disorganized" property is much more probable. Indeed, the tendency of the universe is to "increase its disorganization", an idea that is expressed as the second law of thermodynamics (the law that states that entropy either increases or stays constant, but never diminishes). As we will see later in this chapter, entropy quantifies the number of states with a given property. That is why it is said that entropy is a measure of disorganization.

In a macroscopic system, there is a very large number of elements (for example, in a gas, liquid or solid, N is the number of atoms or molecules in the system). Thus, in a macroscopic system, the fractional width is vanishingly small. For example, at room temperature and pressure, a gas has $\approx 10^{23}$ molecules in 1 cm^3. So, for a macroscopic gas, $N \approx 10^{23}$, thus resulting in a fractional width of the order of 10^{-12}!! So, in a measurement of energy, the error around the most probable energy is of the order of 10^{-10}%. Such a small variation is insignificant in the more fundamental sense that it is not usually detectable; thus, one is physically justified in asserting that the system does have the most probable property.

Notice that, even though Equation (1.9) is only valid for binary systems (where there are only two possible states for each element), the conclusion that the histogram concentrates around the most probable value is general. Indeed, in macroscopic systems, the function describing the probability density is similar to the delta of Dirac function. One example of a non-binary system could be given by replacing the coins by dice. If we agree that the energy is the number facing up, then a macroscopic system of dice with six faces has a most probable energy $U = 3.5 \times N$.

Let us summarize the line of reasoning we have been following: using the examples of coins, we have seen that, if the number N of coins is very large, then the number of states having half of the tails up is overwhelmingly larger than the number of states having other numbers of tails up. Thus, if you toss a macroscopic number of coins, it is virtually guaranteed (within the vanishingly small fractional width) that half of the coins will land tails up. If the number of tails up defines the property of the system, then we can say that the final property is the most probable (and it is $U = N/2$).

This idea that *the final property of a macroscopic system is the most probable* is the foundational idea for the notion of equilibrium in statistical physics. And the reason why this is true is the same as in the example of the coins: the most probable property is overwhelmingly more probable than the others just because there are many more states having the most probable property as compared to the other properties. And we have already hinted at the idea that this most probable property is the one associated with the highest degree of disorganization. Notice, however, an important detail: the whole reasoning assumes that all the states are equally probable. If they are not equally probable, then it is not possible to assert that the most probable property is the one to be observed. Thus, if one tosses a macroscopic number of loaded coins, one will not be surprised if they all land tails up. The assumption that all states are equally probable is the "fundamental assumption of statistical physics".

Let us illustrate the central ideas of statistical physics using another example, now closer to a realistic physical system than just a bunch of coins.

Suppose that you have a gas confined in a gas chamber. Say that there are about 10^{23} molecules of the gas. Suppose that at time t_0 all molecules have been trapped in one corner of the chamber by a magnetic field. Then you switch off the field, thus releasing the molecules. Your job is to study the temporal evolution of the gas and, for definiteness, the property you are interested in is the spatial distribution of the gas. As a marvellous student of physics, you are quite aware that the initial state of the gas is given by the positions and momenta of all the molecules, and that the laws governing the temporal evolution of the initial state are Newton's laws together with Maxwell's equations, the latter giving you the forces applied to the molecules.

So, you know very well what lies ahead of you: you have to solve a set of 10^{23} coupled differential equations (each equation governing the evolution of one of the molecules, and they are coupled because the force on one molecule depends on the position of all the other molecules through Coulomb's law). Of course, you will not be able to solve a set of 10^{23} coupled differential equations. Not you, and not any computer available at the time of the writing of this book. So, Newtonian physics and electrodynamics are helpless in this situation. What shall you do? Well, you can guess. You can assume that all states are equally probable. Then, because the system is macroscopic, you know that the number of states with the most probable property is overwhelmingly larger than the number of states with the other properties. So, a very reasonable and intelligent decision is to give up entirely the daunting task of solving 10^{23} coupled differential equations, and just ask yourself: what is the most probable property? In this example, we said that the property of interest is the spatial distribution of the molecules. In other words: the property is the gas density. It is evident that the most probable property is a uniform distribution of molecules, that is, a uniform density inside the chamber. So, you do not need to solve equations, because you know that the system evolves to a uniform distribution of molecules. And that is true because this is the most "disorganized" property, that is, the property for which there are most states. And you know that, once a uniform distribution is reached, this property is no longer altered (even though the states are changing in time, since the molecules keep moving), because the condition of equilibrium has been reached.

Maybe you are uncomfortable with the idea of classifying a uniform distribution of molecules as the most "disorganized" property, since the notion of uniformity may be associated with organization, and not the other way around. But the physical and popular notions of disorganization are in agreement. The following example may be helpful in clarifying it: suppose you have two friends addicted to shoes. One friend is an organized person, and the other is a disorganized person. If you walk into the room of the organized friend, you expect to find all shoes nicely tidied up in a corner of the room (probably inside a wardrobe). So, the "organized" spatial distribution of shoes is highly heterogeneous (as opposed to homogenous): the shoes are concentrated in a specific corner. If you walk into the room of the disorganized friend, however, you expect to see a homogenous (uniform) spatial distribution of shoes, that is, you expect to see shoes scattered everywhere in the room.

As commented earlier, the second law of thermodynamics is the law of the increase of entropy. Even though this is called a law, in my opinion it is more like a probabilistic observation than a law in the sense of Newton's law, but this is more a matter of philosophical interpretation than of pure physics. As has already been hinted at and will be discussed in detail in the next section, entropy is a measure of the number of states with a given property. Thus, what the second law is essentially affirming is that the system evolves to the property that has most states (that is, the most probable property). In the example of shoes, the room of the disorganized person represents a thermal system with high entropy, since there are many states (here a state is defined by the position of each shoe) giving the property "disorganization". Indeed, if you walk into your disorganized friend's room, and get a blue shoe and fling it somewhere else, your friend will not notice, because this new state still holds the property "a mess". If you do the same experiment in your organized friend's room, she or he will instantly notice the change, because now the property has changed significantly: there is a blue shoe in a position that was previously empty of shoes. The shoes of your organized friend thus represent a system with low entropy, since there is just a small number of states having this property (in fact, just one acceptable state if your organized friend is neurotic).

To conclude this introductory section, it may be important to note that statistical physics is not a fundamental theory in the same sense as quantum mechanics, Newton's laws or electrodynamics. A fundamental law is one that is not entailed by other laws. So, the set of fundamental laws is the minimum set of physical laws from which all the others follow logically. For example, in the realm of classical physics (ignoring relativity for the sake of simplicity), Newton's law and the laws of electrodynamics (Maxwell's equations) are the fundamental laws. Statistical physics, on the other hand, is not fundamental in this sense. In principle, one does not need statistical physics to describe the evolution of the systems: statistical physics is only required for a "practical" reason, that is, when the system is too big to apply the fundamental laws to it.

In the sext section, we will apply our principle that "the system evolves to its most probable property" to a paradigmatic case, which is a system formed by two interacting subsystems. We will begin by expressing our principle in the language of mathematics, which will in turn lead to the notion of thermal equilibrium.

1.2   Thermal equilibrium

Consider an isolated system comprising two interacting macroscopic closed subsystems: system A and system B. System A interacts with B and vice-versa, but there is no interaction with the outside world. By assuming that systems A and B are closed, we are assuming that the number of particles inside each subsystem does not change. So, the interaction is through transfer of energy (here, we are not concerned with the mechanism underlying the energy transfer, but usually the energy transfer is mediated by collision and by exchange of photons).

Suppose that, at the beginning of the interaction, the energy of system A is U_{A0} and the energy of system B is U_{B0}. Since energy is conserved, the energies U_A and U_B at any time obey the relation $U_A + U_B = U_{A0} + U_{B0} = U_0$, where U_0 is the total energy.

The question we want to answer in this section is: what can be affirmed about energies U_A and U_B after the system has reached equilibrium? To answer this question, we are going express in mathematical language the idea that the property in equilibrium is the most probable property.

But before we do that, let us call attention to a few subtleties that may lead to confusion. First of all, notice that we can talk about states of the sub-system A, states of the subsystem B, and states of the total system A + B. The important thing to remember is that the principle of most probable property is applied to the total system A + B, but not to the subsystems A or B. For example, suppose that the subsystems consist of two coins each. The total system then consists of four coins. Thus, if the subsystem A is in the state HH, and the subsystem B is in the state HT, then the state of the total system is HHHT. Furthermore, in this state, the energy of subsystem A is $U_A = 0$, the energy of subsystem B is $U_B = 1$, and the energy of the total system (that is, the total energy) is, of course, $U_0 = U_A + U_B = 1$. But, as the subsystems interact, they can exchange energy. So, if I measure the total system now, and find it to be in the state HHHT, any later measurement can find the system in any state whose total energy is $U_0 = 1$. For example, it could be that, in a few minutes, I check the system again and find to be in the state HHTH, or in the state THHH (notice that $U_A = 0$ in the former and $U_A = 1$ in the latter). Finally, recall that, though the evolution of the system may be deterministic, a macroscopic system is too complicated for us to predict its evolution (the system of four coins is not macroscopic, I am only using it to establish the terminology and explain the logic, but eventually we will consider macroscopic systems). Thus, with a proud heart, we accept our ignorance and apply the fundamental assumption of statistical physics, that is, we just assume that all possible states of the total system are equally probable.

So, continuing with our example of four coins and $U_0 = 1$, we can determine what the possible energies of subsystem A are, and what their corresponding probabilities are. Since we have only two coins in subsystem A, then the possible states of this system (taken in isolation) are HH, HT, TH and TT; so, the possible energies of subsystem A (taken in isolation) are $U_A = 0$, $U_A = 1$ and $U_A = 2$. But, in our example, we have $U_0 = 1$, which means that $U_A = 2$ will never be observed. Thus, the probability of state TT to be found in subsystem A when it is interacting with B is zero if the total energy is $U_0 = 1$. We can work out the probabilities of each state and each possible energy of subsystem A by making a list of all possible states of the total system. Still assuming $U_0 = 1$, we have in total four states of the total system. They are: State 1: HHHT, State 2: HHTH, State 3: HTHH, State 4: THHH (notice that, indeed, the state TT of subsystem A did not appear in the list). In State 1 and State 2, the energy of subsystem A is $U_A = 0$, and in State 3 and State 4 the energy of subsystem A is $U_A = 1$. As expected, no state has energy $U_A = 2$, so the probabilities are 50%, 50% and 0% for $U_A = 0$, $U_A = 1$ and $U_A = 2$, respectively.

It is helpful to use this example to introduce the "multiplicity", which is the number of states having a given energy. We denote the multiplicity by the function $g(U)$. So, if I consider only subsystem A, ignoring interactions with B, then there are four possible states: HH, HT, TH, TT. The energies of these states are, respectively, $U_A = 0$, $U_A = 1$, $U_A = 1$, $U_A = 2$. So, for this system, there is only one state with energy $U_A = 0$. That means that $g_A(0) = 1$, where the subindex A denotes that the multiplicity refers to subsystem A on its own, without considering interactions. Likewise, we have $g_A(1) = 2$, because there are two states with energy $U_A = 1$; and, finally, since there is only one state with energy $U_A = 2$, then $g_A(2) = 1$.

We want to express the multiplicity of the total system, which we call $g_t(U_A)$, in terms of the multiplicities of the subsystems A and B, i.e. $g_A(U_A)$ and $g_B(U_B)$. Let us make very clear what the arguments of these functions mean. First, notice that we have $g_t(U_A)$ instead of $g_t(U_0)$. The subindex A in the argument denotes that this is the energy of the subsystem A. Thus, $g_t(U_A)$ **is the number of states of the total system, in which the subsystem A has energy U_A.** In our example for $U_0 = 1$, two states (States 1 and State 2) have $U_A = 0$, therefore $g_t(0) = 2$. Furthermore, another two states (States 3 and State 4), have energy $U_A = 1$, therefore $g_t(1) = 2$. Finally, no state has energy $U_A = 2$, so $g_t(2) = 0$. So, the function $g_t(U_A)$ is a kind of hybrid, because it ascribes the number of states of the *total* system in terms of the energy of the *subsystem* A. The multiplicities $g_A(U_A)$ and $g_B(U_B)$, on the other hand, refer to the energies of the subsystems taken by themselves, so U_A and U_B are still the energies of the respective subsystems. You may be wondering why g_t was defined in terms of U_A instead of it being defined in terms of the total energy U_0. The reason is that the total energy is fixed (due to energy conservation), which means that, if g_t had been defined in terms of the total energy, it would have been a trivial function (it would be 0 for all energies different than the total energy).

Our job now is to express $g_t(U_A)$ in terms of $g_A(U_A)$ and $g_B(U_B)$. The logic is simple, and it is again best illustrated by means of an example. Let us go back to our good old four states of the total system when $U_0 = 1$. We found that there are two states (State 1 and State 2), in which the energy of the subsystem A is $U_A = 0$, that is, we found that $g_t(0) = 2$. But why is that? First, if we look at the list of possible states of subsystem A, we find that, out of the four possible states of the independent system A, only one state has energy $U_A = 0$ (the state HH). So, we know that $g_A(0) = 1$. But, if the total energy is $U_0 = 1$, and if the subsystem A is in a state with $U_A = 0$, then subsystem B must be in a state having $U_B = 1$. Subsystem B has two coins, so we know that there are two states with $U_B = 1$, that is, HT and TH. That means that the state HH of subsystem A can combine with the states HT and TH of subsystem B, thus forming State 1 (HHHT) and State 2 (HHTH) of the total system, respectively. So, we see that the reason why $g_t(0) = 2$ is that we have one state in subsystem A with energy $U_A = 0$, and two states in subsystem B with the complementary energy $U_B = U_0 - U_A = 1 - 0 = 1$. That is, $g_t(0) = 1 \times 2 = g_A(0) \times g_B(1)$.

If you follow the same reasoning for $g_t(1) = 2$, you will find that there are two states in subsystem A having $U_A = 1$, but only one state in subsystem B having $U_B = U_0 - U_A = 1 - 1 = 0$. So, each state of subsystem A can combine with only one state of system B (thus forming State 3 and State 4 of the total system). That is: $g_t(1) = 2 \times 1 = g_A(1) \times g_B(0)$.

So, generalizing, if there are $g_A(U_A)$ states in subsystem A with energy U_A, and each of these can combine with $g_B(U_0 - U_A)$ states of the subsystem B to form a state of the total system with energy U_0, then the number of states of the total system in which the energy of the subsystem A is U_A is:

$$g_t(U_A) = g_A(U_A)g_B(U_0 - U_A) \tag{1.12}$$

Equation (1.12) is the starting point of our analysis. Remember the question we asked: given two interacting subsystems A and B, what can be affirmed about the energy of the subsystems once equilibrium is reached? To answer this question, we can use the notion that, if the systems are macroscopic, then, after equilibrium is reached, the energy U_A of subsystem A is

Box 2   Another example of multiplicity of a total system

An example showing the construction of the multiplicity of the total system in terms of the product of the multiplicity of the subsystems is shown in Figure B2.1. The example assumes total energy $U_0 = 2$, and that system A has two coins and system B has three coins.

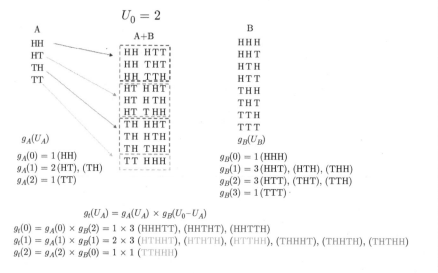

$$g_t(U_A) = g_A(U_A) \times g_B(U_0 - U_A)$$

$g_t(0) = g_A(0) \times g_B(2) = 1 \times 3$ (HHHTT), (HHTHT), (HHTTH)
$g_t(1) = g_A(1) \times g_B(1) = 2 \times 3$ (HTHHT), (HTHTH), (HTTHH), (THHHT), (THHTH), (THTHH)
$g_t(2) = g_A(2) \times g_B(0) = 1 \times 1$ (TTHHH)

Figure B2.1 Multiplicity of the total system in terms of products of the multiplicities of the subsystems A and B, assuming $U_0 = 2$ and two coins in system A and three coins system B. H stands for Heads and T for Tails. The example again assumes that H has zero units of energy and T one unit of energy.

the most probable energy (we can do the analysis for either U_A or U_B, but it does not matter which one we pick because, due to energy conservation, they are not independent of each other). To understand what this means, imagine that there is a list of all possible states of the total system (just like the ones we have been drawing for the systems of coins). These states are composed of combinations of states of subsystem A and states of subsystem B, with the constraint that the total energy is fixed. So, one can count how many of these total states have energy U_A in the subsystem A, just like we have been doing. There will be a value of energy, call it U_f, for which the number of states is the largest. If the systems are macroscopic, and the states of the total system in this list are equally probable, then we know that this most probable energy U_f

is overwhelmingly more probable than the others, thus allowing us to affirm that the energy of the subsystem A after equilibrium is indeed U_f.

Now, we have just seen that U_f is the most probable energy of the subsystem A in the list of possible states of the total system. That means that this is the energy with the largest number of states. But the number of states is given by $g_t(U_A)$. So, we know that U_f is the energy that maximizes $g_t(U_A)$. Furthermore, if the system is macroscopic, we can treat U_A as a continuous variable. We also know from calculus that there is a neat trick to find the maximum of a function: check for which point its first derivative vanishes. That is, we know that U_f is such that:

$$\left.\frac{dg_t}{dU_A}\right|_{(U_A = U_f)} = 0 \tag{1.13}$$

So, the energy $U_A = U_f$ is the energy that maximizes the multiplicity, that is, it is the energy that satisfy Equation (1.13). We can express Equation (1.13) in terms of $g_A(U_A)$ and $g_B(U_0 - U_A)$. From Equation (1.12), we have:

$$\frac{dg_t}{dU_A} = \frac{dg_A(U_A)}{dU_A} g_B(U_0 - U_A) + g_A(U_A)\frac{dg_B(U_0 - U_A)}{dU_A} \tag{1.14}$$

Substituting Equation (1.14) into Equation (1.13):

$$\left[\left.\frac{dg_A(U_A)}{dU_A}\right|_{(U_A = U_f)}\right] g_B(U_0 - U_f) + g_A(U_f)\left[\left.\frac{dg_B(U_0 - U_A)}{dU_A}\right|_{(U_A = U_f)}\right] = 0$$

$$\tag{1.15}$$

To clean up the notation, I will omit the fact that the derivatives are taken at $U_A = U_f$. Thus Equation (1.15) reads:

$$\left[\frac{dg_A(U_A)}{dU_A}\right] g_B(U_0 - U_A) + g_A(U_A)\left[\frac{dg_B(U_0 - U_A)}{dU_A}\right] = 0$$

Using the chain rule and the fact that $U_B = U_0 - U_A$, we can express the last derivative in terms of U_B:

$$\frac{dg_B(U_0 - U_A)}{dU_A} = \frac{dg_B(U_0 - U_A)}{dU_B}\frac{dU_B}{dU_A} = -\frac{dg_B(U_0 - U_A)}{dU_B} = -\frac{dg_B(U_B)}{dU_B}$$

Thus, Equation (1.15) can be written as:

$$\frac{dg_A(U_A)}{dU_A} g_B(U_B) - g_A(U_A)\frac{dg_B(U_B)}{dU_B} = 0 \tag{1.16}$$

Rearranging:

$$\frac{1}{g_A(U_A)} \frac{dg_A(U_A)}{dU_A} = \frac{1}{g_B(U_B)} \frac{dg_B(U_B)}{dU_B} \tag{1.17}$$

Remember that, for a given function $f(x)$:

$$\frac{dln(\ f(x))}{dx} = \frac{df(x)}{dx} \frac{dln(\ f)}{df} = \frac{df(x)}{dx} \frac{1}{f(x)}$$

So, Equation (1.17) can be expressed as:

$$\frac{dln(g_A(U_A))}{dU_A} = \frac{dln(g_B(U_B))}{dU_B} \tag{1.18}$$

To have a cleaner notation, we define the function σ as the natural logarithm of the multiplicity:

$$\sigma = \ ln\,(g) \tag{1.19}$$

With this definition, Equation (1.18) reads:

$$\frac{d\sigma_A}{dU_A} = \frac{d\sigma_B}{dU_B} \tag{1.20}$$

Equation (1.20) is one of the most important results of statistical physics. It tells us that, in equilibrium the derivative of the natural logarithm of the multiplicity with respect to the energy is the same in both subsystems. Notice that this mouthful phrase is just another expression of the probabilistic principle enunciated at the beginning of the previous section. The physics of the interaction between two closed subsystems kind of end in Equation (1.20). This is not, however, the way it is usually expressed. Indeed, though the function σ does not have a specific name, it is used to define the famous function S as:

$$S = k_B\sigma \tag{1.21}$$

where k_B is the famous Boltzmann constant.

As I was saying, even though poor σ has no name, the function S, as defined in Equation (1.21), is immensely famous and goes by the name of **entropy**. Bowing to the weight of fame and tradition, we may re-express Equation (1.20) in terms of the entropy:

$$\frac{dS_A}{dU_A} = \frac{dS_B}{dU_B} \tag{1.22}$$

So, if you are asked the question "What happens when two systems are allowed to interact?", your answer should be that, after equilibrium, the derivative of the entropy with respect to the energy in both subsystems are equal. That is still not a convenient way of expressing the condition of equilibrium. We need a shortcut. For that, we can define T as:

$$\frac{1}{T} = \frac{dS}{dU} \tag{1.23}$$

Now, if you are asked the same question, you can just answer that, after equilibrium, $T_A = T_B$. Of course, T also receives a name. As you have probably guessed, T, as defined in Equation (1.23), is the **temperature**. This is the most fundamental definition of temperature.

This kind of interaction between two subsystems, where only energy is exchanged, is called a thermal interaction (we will soon expand it to include interchange of elements). So, translated into English, Equation (1.22) may be rendered as: *at thermal equilibrium, the two subsystems have the same temperature.* And why is this true? As we have seen, the correct answer is: it is true because it is the most probable thing to happen (see Box 3).

Box 3   Why does energy flow from a system with higher temperature to a system with lower temperature?

Consider again two interacting subsystems A and B, forming a total system A+B. Suppose that they are not in thermal equilibrium, so $T_A \neq T_B$. What is the direction of flow of energy towards equilibrium?

To answer this question, let us consider that a small amount of energy ΔU is exchanged between subsystem A and B. Suppose that the "initial energies" of subsystems A and B – that is, the energies before the exchange of ΔU – are, respectively, U_{A0} and U_{B0}.

If subsystem A lost ΔU to subsystem B, then the "new" energy U_{AN} of subsystem A is:

$$U_{AN} = U_{A0} - \Delta U$$

But if subsystem A lost ΔU to subsystem B, then subsystem B gained ΔU, which entails that the "new energy" U_{BN} of subsystem B is:

$$U_{BN} = U_{B0} + \Delta U$$

With this convention, the direction of flow is determined by the sign of ΔU: on the one hand, if ΔU is positive, then energy flowed from A to B; on the other hand, if ΔU is negative, then energy flowed from B to A.

(Continued)

(Continued)

How can we find the direction of flow of energy or, in other words, the sign of ΔU? To answer this question, we need to inspect what happens to the entropy of the total system (system A + B). Recall that the multiplicity of the total system was defined as a function of the energy of one of the two subsystems (and I chose subsystem A). In the same spirit, we express the total entropy as function of the energy of subsystem A, that is: $S_T(U_A)$. What is $S_T(U_A)$? We saw that, in general, the entropy is defined as $S = k_B \ln g$, where g is the multiplicity. So, it follows that $S_T(U_A) = k_B \ln g_T(U_A)$, where $g_T(U_A)$ is the number of states of the total system having the subsystem A with energy U_A. So, $S_T(U_A)$ is a measure of the number of states of the total system having the subsystem A with energy U_A.

Now, let us define the variation of the entropy ΔS_T as the difference between the entropy after the exchange of ΔU and the entropy before the exchange of ΔU, that is:

$$\Delta S_T(U_A) = S_T(U_{AN}) - S_T(U_{A0})$$

We need to identify the sign of $\Delta S_T(U_A)$ (whether it is positive or negative), but before we do that, let us spell out the meaning of this sign. Suppose that it turns out that $\Delta S_T(U_A) > 0$. What does that mean? Of course, it means that $S_T(U_{AN}) > S_T(U_{A0})$. But, since $S_T(U_A)$ is a measure of the number of states of the total system having subsystem A with energy U_A, the assertion that $S_T(U_{AN}) > S_T(U_{A0})$ is logically equivalent to the assertion that "there are more states of the total system where the energy of subsystem A is U_{AN} than states of the total system where the energy of subsystem A is U_{A0}". Thus, the assertion that $\Delta S_T(U_A) > 0$ is logically equivalent to the assertion that "there are more states with the new energy U_{AN} than states with the initial energy U_{A0}". Likewise, if $\Delta S_T(U_A) < 0$, then there would be less states with the new energy than states with the initial energy.

So, how can we answer the question "what is the direction of the flux of energy"? To the dismay of Rene Descartes, we cannot give a definitive answer. In the spirit of statistical physics, all we can do is to identify the most probable direction of the flux. In other words, all we can do is to humbly ask ourselves: are there more states with energy U_{AN} when $U_{AN} < U_{A0}$ or are there more states with energy U_{AN} when $U_{AN} > U_{A0}$? On the one hand, if there are more states with energy U_{AN} when $U_{AN} < U_{A0}$, then it is more probable that system A will lose energy, so it is more probable that energy will flow from A to B (that is, it is more probable that $\Delta U > 0$). On the other hand, if there are more states with energy U_{AN} when $U_{AN} > U_{A0}$, then it is more probable that energy will flow from B to A (that is, it is more probable that $\Delta U < 0$).

Thus, all we need to do is to identify if there are more states with U_{AN} when $U_{AN} < U_{A0}$ or if there are more states with U_{AN} when $U_{AN} > U_{A0}$. To do that, we need to find the condition that gives a positive sign to $\Delta S_T(U_A) = S_T(U_{AN}) - S_T(U_{A0})$. If, on the one hand, this sign is positive when $U_{AN} < U_{A0}$, then there are more states with U_{AN} when $U_{AN} < U_{A0}$

and, consequently, energy flows from A to B. If, on the other hand, this sign is positive when $U_{AN} > U_{A0}$, then there are more states with U_{AN} when $U_{AN} > U_{A0}$ and, consequently, energy flows from B to A.

To identify the condition that gives a positive $\Delta S_T(U_A)$, we need to express $\Delta S_T(U_A)$ in terms of ΔS_A and ΔS_B. We already know that:

$$g_t(U_A) = g_A(U_A)g_B(U_B)$$

Taking the natural logarithm on both sides:

$$\ln g_t(U_A) = \ln\left[g_A(U_A)g_B(U_B)\right] = \ln g_A(U_A) + \ln g_B(U_B)$$

Multiplying both sides by k_B, it follows that:

$$S_T(U_A) = S_A(U_A) + S_B(U_B)$$

From which it follows immediately that

$$\Delta S_T(U_A) = \Delta S_A(U_A) + \Delta S_B(U_B)$$

Since

$$\frac{1}{T} \approx \frac{\Delta S}{\Delta U}$$

Then:

$$\Delta S \approx \frac{\Delta U}{T}$$

Where the equality holds for infinitesimal variations. Keeping the equal sign, we thus have:

$$\Delta S_T(U_A) = \frac{\Delta U_A}{T_A} + \frac{\Delta U_B}{T_B}$$

But $\Delta U_A = U_{AN} - U_{A0} = -\Delta U$ and $\Delta U_B = U_{BN} - U_{B0} = +\Delta U$. Thus:

$$\Delta S_T(U_A) = \frac{-\Delta U}{T_A} + \frac{\Delta U}{T_B} = \Delta U\left(\frac{1}{T_B} - \frac{1}{T_A}\right)$$

By inspection of the equation above, we conclude that, if $T_A > T_B$, then $\Delta S_T(U_A) > 0$ when $\Delta U > 0$.

Let us spell out this terse mathematical conclusion in English. We have concluded that, when subsystem A is hotter than subsystem B ($T_A > T_B$), it is more probable ($\Delta S_T(U_A) > 0$) that energy flows from subsystem A to subsystem B ($\Delta U > 0$).

If, on the other hand, $T_A < T_B$, then $\Delta S_T(U_A) > 0$ when $\Delta U < 0$.

Again, let us spell it out in English. We have concluded that, when subsystem B is hotter than subsystem A ($T_B > T_A$), it is more probable ($\Delta S_T(U_A) > 0$) that energy flows from subsystem B to subsystem A ($\Delta U < 0$).

(Continued)

(Continued)

> In both cases, we concluded that it is more probable that energy goes from the hotter system to the colder system.
>
> So, why does energy flows from the hotter system to the colder system, and not the other way around? The honest answer is: I don't know! All I know is that it is more probable that energy flows from the hotter system to the colder system. If you like to talk fancy, you can say the same thing by stating that "according to the second law of thermodynamics, the entropy of the total system cannot decrease; thus energy flows from the hotter system to the colder system because this is the process that increases the entropy (that is, that gives $\Delta S_T(U_A) > 0$)". But never forget that this fancy talk just means that it is more probable that hotter heats the cold than cold heats the hot, and that is all we can know about it.

In this section, we concluded that, when two subsystems interact by exchanging energy, equilibrium is reached when the temperatures equalize. We saw that this is just the condition that gives the most probable property. Even though this is a result of tremendous importance in statistical physics, it assumes that both systems are macroscopic, which is not always the case: often, one subsystem is microscopic, and the other is macroscopic. Furthermore, we might need more detailed information about the probabilities of the properties and states of the subsystems. This is the subject of the next section.

1.3   The partition function – part I

Consider again the case of two closed and interacting subsystems, A and B, forming a total system A + B, as we treated in the previous section. Often, we are interested in studying one of the subsystems, say, system A, when it is in contact with the other, about which we have some information. So, for example, our subsystem A can be our semiconductor, and subsystem B can be the laboratory where the semiconductor is in. And, unless our economic situation is so dire that we cannot afford a thermometer, we will have some information about our laboratory, like its temperature.

Our job now is to obtain an expression for the probability of finding a given state of subsystem A in terms of parameters of subsystem B. Our system A, that is, our system of interest, can be either macroscopic or microscopic, but we will assume that system B is macroscopic, and I will make it very explicit where we are using this assumption.

So, suppose I check the list of possible states of system A, and pick out a specific one, say, state β. We want to know the probability that, in a measurement of the total state (system A + B) of the system, we find the subsystem A in the state β. Please read this last sentence one more time to make sure you understood it.

To see how we can find this probability, consider again Equation (1.12). We have seen that $g_t(U_A) = g_A(U_A)g_B(U_0 - U_A)$ because, if there are $g_A(U_A)$ states in system A (taken in isolation), then each of these states can combine with each of the $g_B(U_0 - U_A)$ states of system B with the complementary energy to form a state of the total system where the energy of the subsystem A is U_A. But now we are not interested in all states of system A that have energy U_A; instead, we are interested in a specific state, which we denoted as β. How many times does the state β of system A appear in the list of the total states of the total system? Well, if the energy of the state β is ε_β, then we know that this state β can combine with any state of the system B that has the complementary energy, that is, with any state of system B that has energy $U_0 - \varepsilon_\beta$. But the number of states of system B having energy $U_0 - \varepsilon_\beta$ is $g_B(U_0 - \varepsilon_\beta)$. So, state β of system A can combine with $g_B(U_0 - \varepsilon_\beta)$ states of system B. Consequently, state β of system A appears $g_B(U_0 - \varepsilon_\beta)$ times in the list of states of the total system. **So, there will be $g_B(U_0 - \varepsilon_\beta)$ states of the total system where the state of the subsystem A is β.**

To make sure this logic is driven home, let us again resort to the example of two coins in system A and two coins in system B. But now let us name the states of system A. Let us call α the state HH, κ the state HT, γ the state TH and ω the state TT. We assume again that $U_0 = 1$. We have seen that, in this case, there are four states of the total system: State 1: HHHT, State 2: HHTH, State 3: HTHH, State 4: THHH. So, we can see that the state α (that is, HH) of system A appears twice in the list, that is, it appears in State 1 and State 2. And state α appears twice because there are two states of system B (HT and TH) that can combine with state α of system A within the constraint that $U_0 = 1$. Indeed, since the energy of state α is $\varepsilon_\alpha = 0$ (as there are no tails up), a state of system B combining with α must have energy equal to 1. But there are two states of B with energy 1 (HT and TH). In terms of multiplicity, this means that $g_B(U_0 - \varepsilon_\alpha) = g_B(1 - 0) = g_B(1) = 2$. In conclusion, state α appears twice in the list of total states because $g_B(U_0 - \varepsilon_\alpha) = 2$. State κ, on the other hand, has energy $\varepsilon_\kappa = 1$. Now, the state of system B must have energy equal to 0; but there is only one state of system B with energy equal to 0 (HH). So, state κ appears only once (it appears only in State 3), because $g_B(U_0 - \varepsilon_\kappa) = g_B(1 - 1) = g_B(0) = 1$.

Box 4   Another example of probability of finding a given state of system A in terms of the multiplicity of system B

Another example showing the construction of the probability $P(\beta)$ of finding a state β of system A in terms of the multiplicity of system B is shown in Figure B4.1. The example assumes total energy $U_0 = 2$, and that system A has two coins and system B has three coins.

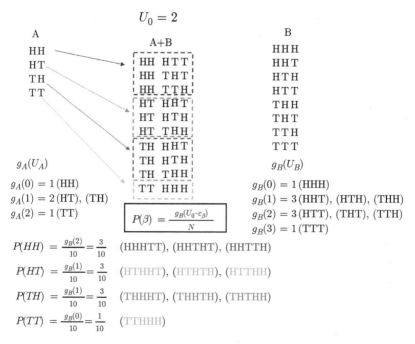

Figure B4.1 Probability $P(\beta)$ of finding a state β of system A in terms of the multiplicity of system B, assuming $U_0 = 2$ and two coins in system A and three coins in system B. H stands for Heads and T for Tails. The example again assumes that H has zero units of energy and T has one unit of energy.

Following this logic, if there are N states of the total system A + B, since a state β of system A appears $g_B(U_0 - \varepsilon_\beta)$ times in this list of N possible states (β here is any state whatsoever), so the probability $P(\beta)$ of finding system A in the state β is:

$$P(\beta) = \frac{g_B(U_0 - \varepsilon_\beta)}{N} \qquad (1.24)$$

Notice, once again, that we are assuming that the N possible states of the total system are equally probable (the fundamental assumption of statistical physics); otherwise Equation (1.24) would not be true.

Equation (1.24) expresses the probability of finding a state β of system A in terms of parameters of the other system, that is, of system B, since it is the multiplicity of system B that appears on the right-hand side of Equation (1.24). So, that is kind of good, because in practice we have some information about system B, like its temperature. But Equation (1.24) is not very useful. One problem is that it depends on N, which is the number of states of the total system. In our example of four coins with $U_0 = 1$, we can readily identify that we have four states of the total system (States 1,2,3,4), so it is easily found that $N = 4$ in that example. But, if the system is macroscopic, it is almost always practically impossible to find N. Thus, let us get rid of N by dividing the probability of finding a state β by the probability of finding another state α (here α is any state whatsoever, not necessarily the α state of the previous example):

$$\frac{P(\beta)}{P(\alpha)} = \frac{g_B\left(U_0 - \varepsilon_\beta\right)}{N} \frac{N}{g_B\left(U_0 - \varepsilon_\alpha\right)} = \frac{g_B\left(U_0 - \varepsilon_\beta\right)}{g_B\left(U_0 - \varepsilon_\alpha\right)} \tag{1.25}$$

This form is better, because now we do not need to worry about N. But Equation (1.25) is still not very useful, because it requires knowledge of the multiplicities of each energy, which may be impossible to calculate or measure in a macroscopic system. What really is easy to know is the temperature, so it would be nice if we could express the ratio of Equation (1.25) in terms of the temperature. We know that, by definition, the inverse of the temperature is the derivative of the entropy (Equation (1.23)), so it may be useful to recast Equation (1.25) in terms of the entropy, which can be easily done using Equation (1.19) and Equation (1.21). Thus:

$$\frac{P(\beta)}{P(\alpha)} = \frac{\exp\left[\sigma_B\left(U_0 - \varepsilon_\beta\right)\right]}{\exp\left[\sigma_B\left(U_0 - \varepsilon_\alpha\right)\right]} = \frac{\exp\left[\frac{S_B\left(U_0 - \varepsilon_\beta\right)}{k_B}\right]}{\exp\left[\frac{S_B\left(U_0 - \varepsilon_\alpha\right)}{k_B}\right]}$$

The expression above is valid for both macroscopic and microscopic systems (or any combination of them). But remember that system A is our system of interest and system B is a big system in contact with system A. The technical name for a "big system" is "reservoir". So, our main interest is in finding these probabilities for system A when it is in thermal contact with a reservoir. As a reminder that B is a reservoir, I will replace the subindex B by R. Doing that, and moving the exponential in the denominator up to the numerator, the expression above reads:

$$\frac{P(\beta)}{P(\alpha)} = \exp\left[\frac{S_R(U_0 - \varepsilon_\beta) - S_R(U_0 - \varepsilon_\alpha)}{k_B}\right] \qquad (1.26)$$

Now comes the magic. If system A is in contact with a much bigger system, that is, with a reservoir, then the energies of the total system are typically much higher than the energies of system A. In our notation, that means that $U_0 \gg \varepsilon$, where ε is the energy of a state of A. This condition motivates an expansion of the entropy in a Taylor series around the point U_0:

$$S_R(U_0 - \varepsilon) = S_R(U_0) - \varepsilon\frac{dS_R}{dU} + \frac{\varepsilon^2}{2}\frac{d^2 S}{dU^2} + \dots \qquad (1.27)$$

But we know that the first derivative of the entropy is the inverse of the temperature, so the second term of the Taylor expansion is proportional to the inverse of the temperature. The higher order terms, however, involve derivatives of the temperature. These higher order terms are very small, since they depend on variations of the temperature of the reservoir. For example, if your system A is a piece of semiconductor and R is the laboratory, then the higher order terms depend on how much the temperature of the laboratory changes when you bring in your semiconductor. Since, by definition, the reservoir is much bigger than system A, variations of its temperature can be dismissed, which allows us to retain only the first two terms of the Taylor series (mathematically, we are truncating the Taylor series in its first order term because ε is small, so $\varepsilon \gg \varepsilon^2$, or, if you prefer, we are treating $S_R(U)$ as a line around the point U_0). Thus:

$$S_R(U_0 - \varepsilon) \cong S_R(U_0) - \varepsilon\frac{dS_R}{dU} = S_R(U_0) - \varepsilon\frac{1}{T} \qquad (1.28)$$

Notice that T is, strictly speaking, the temperature of the reservoir. But if system A is in thermal equilibrium with the reservoir, then it is also the temperature of system A.

Using the Taylor expansion with only the first two terms retained, we get:

$$S_R(U_0 - \varepsilon_\beta) - S_R(U_0 - \varepsilon_\alpha) \approx \left[S_R(U_0) - \varepsilon_\beta\frac{1}{T}\right] - \left[S_R(U_0) - \varepsilon_\alpha\frac{1}{T}\right]$$

$$= -\frac{\varepsilon_\beta - \varepsilon_\alpha}{T} \qquad (1.29)$$

Substituting Equation (1.29) into Equation (1.26):

$$\frac{P(\beta)}{P(\alpha)} = \exp\left[-\frac{\varepsilon_\beta - \varepsilon_\alpha}{k_B T}\right] = \frac{\exp\left[-\frac{\varepsilon_\beta}{k_B T}\right]}{\exp\left[-\frac{\varepsilon_\alpha}{k_B T}\right]} \tag{1.30}$$

Equation (1.30) is a piece of beauty. Let us pause a bit to appreciate it. First, remember what kind of information we used to get to this result. The only law of physics that we invoked was the law of conservation of energy. We also used the fundamental assumption of statistical physics that all states are equally probable. And, finally, we assumed that R is a big system, that is, a reservoir. With just these three pieces of information, we obtained the ratio of the probabilities of finding states of the system A, which involves only the energy of the states, and the temperature of the reservoir. Notice that we did not make any detailed assumption about the systems. Indeed, we do not know what A is, if it is a solid, a liquid or a gas. We also do not know if R is the laboratory, or a big volume of a liquid, or maybe a part of the atmosphere. Because we did not make any of these assumptions, Equation (1.30) is valid for any system, as long as it is in thermal equilibrium with a reservoir.

But this masterpiece still needs a final touch. Indeed, it is more convenient to find an expression for the absolute value of the probability, instead of just relative values. But it is not difficult to convert the ratio of probabilities to absolute values. First, notice that, since Equation (1.30) applies to any pair of states, it follows that the absolute probability must be a proportion of the exponential factor. Thus, calling this proportion $1/Z$, the absolute probability of finding a state c must be of the form:

$$P(c) = \frac{\exp\left[-\frac{\varepsilon_c}{k_B T}\right]}{Z} \tag{1.31}$$

To find Z, we can invoke the fact that the sum of the probabilities over all states must be equal to 1:

$$\sum_c P(c) = 1 \tag{1.32}$$

It is important to emphasize that this sum is over all states of system A, not over the states of the total system. Thus, c is a given state of system A. In the example where system A is formed by two coins, the sum is over all four possible states of system A, that is $P(\alpha) + P(\kappa) + P(\gamma) + P(\omega) = 1$, where, as defined earlier in the text α, κ, γ and ω are the states HH, HT, TH and TT, respectively.

Substituting Equation (1.31) into Equation (1.32):

$$\sum_c P(c) = \frac{\sum_c \exp\left[-\frac{\varepsilon_c}{k_B T}\right]}{Z} = 1 \qquad (1.33)$$

Thus:

$$Z = \sum_c \exp\left[-\frac{\varepsilon_c}{k_B T}\right] \qquad (1.34)$$

Notice that the normalization parameter Z is really a function of the temperature, so it is safer to express this dependence on the temperature explicitly. Thus:

$$Z(T) = \sum_c \exp\left[-\frac{\varepsilon_c}{k_B T}\right] \qquad (1.35)$$

This function is extremely important in statistical physics and goes by the name of "partition function". I will show you a classic example of application of this function in section 1.6, and if you want you can read section 1.6 now and then come back to section 1.4.

Though extremely important and useful, the partition function does not consider exchange of elements (we have been considering that subsystem A and B can exchange energy, but we have not considered the possibility of exchange of elements yet), which is an important process in semiconductor devices. So, we still need to extend it, and we will do that in the section labelled as part II of the partition function, where we consider systems that can also change elements, that is, systems in diffusive contact. There, we will see that the partition function also depends on the chemical potential, which is the topic of the next section.

1.4   Diffusive equilibrium and the chemical potential

The theory developed so far is based on energy conservation and on the fundamental assumption of statistical physics, which asserts that all states (of the total system) are equally probable. But we also assumed that the systems do not exchange elements (often we call the elements "particles"). However, we often find situations involving exchange of elements. For example, we could have two interacting gases. We could envisage that each has a different concentration of molecules, and that we open a compartment between them. In this case, we know that molecules diffusive from one system to the other, until equilibrium is reached. In the previous section, where we considered that only energy was exchanged, we found that equilibrium was reached when the two temperatures equalize. Now we need to identify what the conditions of equilibrium are, when not only energy, but also elements, are exchanged. Is there also a quantity whose equalization across the system characterizes diffusive equilibrium? As it turns out, there is: it is the chemical potential. To identify this condition, we can follow a similar method to that of the previous section. The main difference is that, while before we considered only one property (energy), now we need to include a second property in the story: the number of elements.

Before we move on, let us see how this new property turns up in the example of the coins. In the previous sections, the number of coins in each subsystem was fixed (since we assumed that the subsystems did not exchange particles). But, now, the number of coins can change. So, for example, if we start again with two coins in subsystem A and two coins in subsystem B, but let the subsystems exchange both energy and coins, then we need to include the possibilities of different numbers of coins in the list of possible states. Thus, when listing the possible states of system A, we have to consider states with all possible numbers of coins. For example, we would have the state H, which is a state with the properties "one element, and zero units of energy". We would also have the state T, which is a state with the properties "one element, and one unit of energy". Likewise, we can have the state HH, which is a state with the properties "two elements, and zero units of energy". Or the state HHTHT, which is a state with the properties "five elements and two units of energy". And so on.

It is quite straightforward to include this second property, "number of elements", in the analysis of section 1.2. The reasoning is essentially the same as in section 1.2, but now the multiplicity must be a function of two variables: the energy U and the number of elements N (notice that here the meaning of N is not the same as it was in Equation (1.24), sorry for that). Then we follow the same idea: look for the properties that maximize the multiplicity. Since the multiplicity g is now a function of two variables, the requirement is that the partial derivatives with respect to each of these variables vanish. The partial derivative with respect to U leads to Equation (1.22) in the same way as before. Physically, this means that thermal equilibrium is still reached when diffusive contact is considered. The partial derivative with respect to the number of elements N follows in an analogous way. The only difference worth mentioning is that, on the one hand, energy conservation is invoked in the derivation of the thermal equilibrium condition (Equation (1.22)), while, on the other hand, conservation of elements must be invoked in the derivation of the diffusive equilibrium condition. This means that, when taking the partial derivative of g with respect to N, one needs to consider that the total number of elements (of the total system) does not change. Thus, one reaches a conclusion analogous to Equation (1.22), but with the number of particles N in the place of U.

Putting both results together, we find that the two conditions that maximize the multiplicity are:

$$\left(\frac{\partial S_A}{\partial U_A}\right)_{N_A} = \left(\frac{\partial S_B}{\partial U_B}\right)_{N_B} \quad and \quad \left(\frac{\partial S_A}{\partial N_A}\right)_{U_A} = \left(\frac{\partial S_B}{\partial N_B}\right)_{U_B} \tag{1.36}$$

The symbols N and U added to the right-hand corner of each partial derivative are reminders that each partial derivative is taken with the other variable fixed. So, the partial derivative with respect to the energy is taken with the number of elements fixed, and vice-versa. Thus, we are treating the energy and number of elements as free variables. The identification of the free variables plays an important role in thermodynamics, but we cannot treat this issue here (and it is not deadly important for our purposes). So, from now on, to avoid clutter, I will omit the symbols on the right-hand corner. Notice that, since the multiplicity is now a function of U and N, so is the entropy.

If we were kind enough to give a name to the partial derivatives that appear in the first equality of Equation (1.36) (remember that the inverse of these derivatives is called "temperature"), it is only fair that we also give a name to the derivatives involved in the second equality. After all, they express this new condition of equilibrium that characterizes systems that can exchange not only energy, but also elements. The convention is to define the parameter μ as:

$$\mu = -T\frac{\partial S}{\partial N} \tag{1.37}$$

The parameter μ, as defined in Equation (1.37), is the "chemical potential". Notice that the chemical potential is a function of the temperature T.

The easiest way to get an intuition about the role played by the chemical potential is to compare it with the temperature. Notice that the chemical potential is defined in terms of the derivative of the entropy with respect to the number of particles N, analogously to the definition of the temperature in terms of the derivative of the entropy with respect to the energy U. Consequently, we can use our intuitive knowledge about the relationship between temperature and energy to gain an intuition about the relationship between chemical potential and number of elements. We know that temperature defines the direction of energy flux: energy (in the form of heat) flows from systems of higher temperature to systems of lower temperature (see Box 3). Likewise, the chemical potential defines the flow of elements: **elements flow from the system of higher chemical potential to the system of lower chemical potential.** (Why is that? It is for the same reason as explained in Box 3: the total entropy increases as elements flow from the system with higher chemical potential to the systems with lower potential, which essentially means that it is more probable that elements flow from the system with higher chemical potential to the system with lower chemical potential, than it is for the inverse process to happen.)

To summarize the analogy: thermal equilibrium is characterized by constant temperature (the temperature is the same in both systems); likewise, diffusive equilibrium is characterized by constant chemical potential (the chemical potential is the same in both systems). If the systems are in thermal and diffusive equilibria at the same time, then the temperature and chemical potential are the same in both systems. Furthermore, we know that thermal equilibrium is reached through exchange of energy. Likewise, diffusive equilibrium is reached through exchange of elements. Also, in the same way that it is the temperature that decides the direction of flow of energy (from the hotter to the colder), it is the chemical potential that decides the direction of flow of elements: elements diffuse from the system with higher chemical potential to the system with lower chemical potential.

At this stage, you may be wondering if the chemical potential is related to the concentration of elements. They are intimately related, but they are not the same thing (in Chapter 2 we will find an explicit relation between them). For one thing, the chemical potential has units of energy. One of the reasons why temperature is multiplying the derivative of the entropy in the definition of the chemical potential (Equation (1.37)) is precisely to give it units of energy (entropy times temperature has units of energy). This is motivated by the fact that the chemical potential plays a similar role to other potentials in physics. Thus, as is well known, a difference in potential energy results in movement of elements (for example, a difference in gravitational potential energy causes the water current in a river; and a difference in electrostatic potential energy causes the electric current in a resistor). This kind of movement that results from a difference in potential energy is called "drift current". Likewise, differences in chemical potential also make things move, but now through a different type of movement (or, more formally, a different type of transport), which characterizes a "diffusion current". More details about these two types of currents are given in section 1.10.

To gain more insights into the physical meaning of the chemical potential, we can express the variation of entropy in terms of variations of energy and number of elements. From calculus, we know that an infinitesimal variation in a function of two variables, here $S(U, N)$, can be expressed as:

$$\Delta S = \frac{\partial S}{\partial U}\Delta U + \frac{\partial S}{\partial N}\Delta N \qquad (1.38)$$

Notice that the equation above comes directly from the laws of derivatives of functions of two variables. The physical content comes in when we express the partial derivatives in terms of temperature and chemical potential:

$$\Delta S = \frac{1}{T}\Delta U - \frac{\mu}{T}\Delta N \qquad (1.39)$$

Rearranging the equation above, the variation in energy can be expressed as:

$$\Delta U = T\Delta S + \mu\Delta N \qquad (1.40)$$

There are two important features to be noticed in Equation (1.40).

First, it confirms that indeed the chemical potential has units of energy, since the left-hand side of the equation has units of energy and ΔN has no units (it is just the variation in the number of elements).

Second, notice that this equation expresses the energy as the contribution of two terms: $T\Delta S$ and $\mu\Delta N$. The first term is the contribution of "heat" and the second term is the contribution of the "free energy". The following thought experiment offers an intuitive way of understanding the meaning of these two terms: suppose that you have a closed recipient with a gas inside it. The recipient is sealed, and it is in thermal equilibrium with a reservoir at the temperature T. So, the gas has the same temperature T as the reservoir, but its chemical potential μ is not the same as the reservoir's, because there is no diffusive contact in this thought experiment – the recipient is closed. Suppose further that there are no potential energies involved in this problem, so the energy of the gas is due solely to the kinetic energies of the molecules inside it. The total energy of the gas is U.

Now, suppose that you open a small door in the recipient, and throw ΔN molecules inside it (and you left the door open for a very short time, so as not to let any molecule escape). Say that each of the new molecules has a kinetic energy E_K. So, the energy of the gas increased by an amount $\Delta U = E_K\Delta N$. If, immediately after closing the door, you also isolate the gas from the universe (by placing it in an awesome thermal bottle), then the new energy of the gas will be $U + \Delta U = U + E_K\Delta N$, forever and ever. Nothing very mysterious. But, if you do not isolate the gas, then part of this energy will be lost to the reservoir, because the new molecules changed the temperature T of the gas. The gas got hotter, or colder (depending on E_K), so it will exchange energy with the reservoir until it reaches thermal equilibrium again. This energy that is exchanged to reach thermal equilibrium is the $T\Delta S$ bit in Equation (1.40). The energy that remains after thermal equilibrium is reached is the $\mu\Delta N$

bit. Thus, $\mu \Delta N$ is called the "free energy", because it is the energy that remained after thermal equilibrium was reached again (interestingly, notice that the free energy depends on the number of molecules ΔN that you injected into the gas, and on the chemical potential of the gas before the injection, but it does not depend on the energy E_K of the molecules that were injected). In the Appendix, it is offered a geometrical proof of the assertion that "$T \Delta S$ is lost (or gained) to the environment and $\mu \Delta N$ is the energy that remained after thermal equilibrium is reached".

Now we need to consider how the chemical potential interacts with other potentials. In the context of semiconductor physics, we are interested in the interaction with the electrostatic potential. However, it is instructive to first consider the interaction with the gravitational potential. To do so, we shall resort to another thought experiment (Einstein famously loved thought experiments, so we can say that at least that we have in common with the great man).

First, consider again the classic situation of two interacting gases, as shown in Figure 1.5a. We already know that, when the two gases interact, the chemical potential on both sides equalize (the concentration is the same on both sides). After equilibrium is reached, the net flux of elements (i.e., the diffusion current) is zero.

Now, what happens if we place one of these compartments in a region of higher potential energy, as shown in Figure 1.5b? Obviously, the elements in the upper compartment have a higher gravitational potential energy than the elements in the other compartment. This difference in potential energy causes a downward flux of elements, i.e., it causes a downward drift current. Since the upper compartment is losing elements due to the drift current, its concentration, and hence chemical potential, is reduced. Meanwhile, the concentration and chemical potential in the other compartment increase, as it is receiving elements. So, the drift current causes an imbalance between the chemical potentials of the upper and lower systems. This imbalance results in an upwards diffusion current (the current is upwards because it is the compartment at the bottom that has the higher chemical potential). Equilibrium is reached when the *upward diffusion* current cancels the *downward drift* current, that is, when the **total** potential (chemical + gravitational) is the same in both compartments (notice that, at equilibrium, neither the chemical nor the gravitational potentials by themselves are equalized across the system).

Before equilibrium is reached, the net flux is decided by the total potential: if the total potential is higher in the upper compartment, then there will be a net downward current (the drift current will be higher than the diffusion current); if the total potential is higher in the lower compartment, then there will be a net upward current (the diffusion current will be higher than the drift). So, it is the total potential that decides the net flux of elements. And, at equilibrium, the total potential is the same.

This is an important example, so let us ponder about it a little bit. Using only the requirement that the entropy is maximized, we had reached the conclusion that the chemical potential, given by Equation (1.37), is constant across the system at diffusive equilibrium. This is the situation exemplified in Figure 1.5a. But, in Figure 1.5b, the chemical potential is no longer constant, even though the net flux is zero. How can the example of Figure 1.5b be reconciled with the general result that the derivative of the entropy with respect to the number of elements (that is, the chemical potential) has to be constant at equilibrium (remember that this is just the most probable condition, that is, the condition that maximizes the entropy, so it is quite general and it should be also applicable to Figure 1.5b)?

It turns out that **the potential that appears in Equation (1.37) is always the TOTAL potential** (chemical + gravitational + electrostatic +....). Thus, Equation (1.37) describes both the situation of Figure 1.5a and the situation of Figure 1.5b. The difference is that, in Figure 1.5a, the total potential is just the chemical potential, whereas in Figure 1.5b the total potential is the chemical potential plus the gravitational potential.

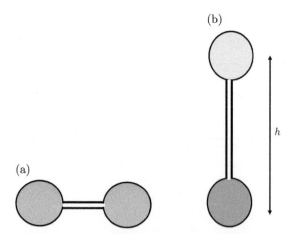

Figure 1.5 Relationship between chemical and gravitational potential. a) Two compartments in diffusive equilibrium have the same chemical potential. b) Now one of the compartments is placed at a higher gravitational potential. The gravitational potential induces a downward drift current, thus sending elements to the lower compartment, which increases its concentration. The higher concentration in the lower compartment results in a higher chemical potential, which induces an upward diffusion current. Equilibrium is now reached when the total potential (gravitational + chemical) is the same in both compartments. At equilibrium, the upwards diffusion current cancels the downward drift current, so that the total current is zero.

To shows this a bit more formally, let us ascribe some symbols to these potentials. The chemical potential will be denoted by the symbol μ_q, the gravitational potential will be denoted by the symbol μ_g, and the total potential will be denoted by the symbol μ_t. Now, consider again Equation (1.40). This equation describes the change of energy in the system when ΔN elements are brought in or out. In the example of Figure 1.5a, the total potential is just the chemical potential. Therefore, for the example of Figure 1.5a, we can write:

$$\Delta U = T\Delta S + \mu_q \Delta N$$

for Figure 1.5a

Rearranging the equation above to isolate the entropy, we find:

$$\Delta S = \frac{\Delta U}{T} - \frac{\mu_q}{T}\Delta N$$

But, from calculus we know that:

$$\Delta S = \frac{\partial S}{\partial U}\Delta U + \frac{\partial S}{\partial N}\Delta N$$

Therefore, we can conclude that:

$$\mu_t = \mu_q = -T\frac{\partial S}{\partial N}$$

in Figure 1.5a

Just as before (Equation (1.37)).

Now, let us repeat this exercise for the example of Figure 1.5b. To adapt it to the situation of Figure 1.5b, notice that ΔU in Equation (1.40) is the energy difference in the system when ΔN elements are brought in or out. If each of these elements adds an additional potential energy of mgh (where m is the mass, g is the gravitational acceleration and h is the height), then ΔU becomes:

$$\Delta U = T\Delta S + \mu_q \Delta N + mgh\Delta N = T\Delta S + \left(\mu_q + mgh\right)\Delta N$$
$$= T\Delta S + \left(\mu_q + \mu_g\right)\Delta N$$

in Figure 1.5b

where we can readily identify that $\mu_g = mgh$.

So, because the chemical and potential energies contributions are both proportional to ΔN, they appear together, forming the total potential. If you follow the same steps as were done for the example of Figure 1.5a, you can easily reach the conclusion that:

$$\mu_q + \mu_g = -T\frac{\partial S}{\partial N}$$

in Figure 1.5b

Thus, if any external potential is included in the systems in diffusive contact, equilibrium is reached when the TOTAL potential is equalized across the systems. This is precisely the result that we expected from the qualitative analysis of Figure 1.5.

Notice that we could have had more than one element. If so, each element would have its own chemical potential, and equilibrium would be reached for each kind of element. The entropy would thus be a function of the energy and of the numbers of each element. For example, if we had two different kinds of elements, say oxygen and hydrogen, then the entropy would be a function of the energy U, the number of oxygen molecules N_1, and the number of hydrogen molecules N_2.

Now we turn our attention to another example, this time more representative of the actual physics of a semiconductor **p-n** junction. Consider two situations, represented in Figure 1.6a and Figure 1.6b. In both cases, the total system consists of two subsystems: subsystem A and subsystem B (they can also be considered in thermal contact with a reservoir, which sets their temperature). The systems are connected by a small pipe. Initially, the systems are identical.

Now, imagine that you raised the chemical potential of free electrons (that is, electrons that are free to move around and thus can contribute to current) in system B, as shown in Figure 1.6a. You can do that by increasing the concentration of free electrons. Notice that you do not need to increase the net charge of system B to do that: you can just increase the concentration of free electrons while keeping the net charge equal zero (more details on how to increase the free electron concentration keeping the net charge zero are given in Chapter 2). So, let us imagine that the chemical potential of free electrons in system B has been increased by an amount μ_0, but that it remains a neutral system (zero net charge). Let the chemical potential of system A be μ_A. So, the chemical potential of system B is $\mu_A + \mu_0$. What is going to happen now that the chemical potential of B is higher? Evidently, there will be diffusion of electrons from B to A, so there will be an electric **diffusion** current from A to B (remember that, thanks to Benjamin Franklin's unfortunate choice of ascribing negative charge to electrons, the electric current is in the opposite direction of the flux of electrons). To facilitate the thought experiment, we assume that the resistance outside the pipe can be neglected. If there is an electric current in the pipe (due to the diffusion current), then there is a potential difference across the pipe. Let us imagine that you measure this difference with a voltmeter in the beginning of the process. Since the electric current is flowing from A to B, then if you plugged the positive cable of the voltmeter on the A side and the negative on the B side, you would measure a positive value, and we call it V_0, as shown in Figure 1.6a. Next, you magically freeze the process, until further notice.

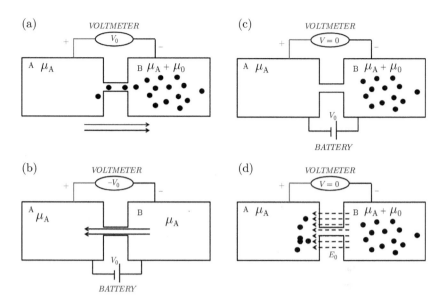

Figure 1.6 Four different physical situations used to illustrate the physical meaning of the electrochemical potential. a) system B has a higher chemical potential than system A, thus inducing an electric diffusion current and hence a potential difference V_0. b) The chemical potential is the same in both systems, but now we added an external battery with a voltage $-V_0$, thus inducing a drift current with the same magnitude, but opposite direction, of the current of situation a). c) Now we have a difference in both chemical and electrostatic potentials, but they cancel each other out and the net current is zero, since the total potential difference is zero. d) we go back to the situation of a), but now let the diffusion current flow: as it flows, charges accumulate on both sides of the systems, thus creating an "internal capacitor", which sets up an internal electric field. The internal electric field causes a drift current in the opposite direction of the diffusion current, and equilibrium is reached when both currents cancel each other, that is, when the net current is zero. At equilibrium, the total potential (chemical + electrostatic) is the same in both systems, but neither the chemical potential by itself, nor the electrostatic potential by itself, are the same in both systems. The total potential is the "electrochemical potential".

Now that we measured V_0 and magically froze the system, let us forget about it and imagine a second experiment, as represented in Figure 1.6b. Now we let the chemical potentials on both sides be the same, but we plug a battery to set up a potential difference of $-V_0$ across the pipe. The battery will induce a **drift** current (contrasting with the previous experiment, that induced a diffusion current), but now from B to A, because we connected the higher potential of the battery at B.

Now we combine the systems of Figure 1.6a and Figure 1.6b. Now, we have both a higher chemical potential in B, but also the battery, as shown in Figure 1.6c. We have seen that the chemical potential sets up a voltage V_0, while the battery adds a voltage $-V_0$. So, the total voltage of the system in Figure 1.6c is zero, which entails that the net current is also zero. But this is happening in a situation where neither the chemical potential nor the electrostatic potential are the same, analogously to the gravitational situation of Figure 1.5b. Indeed, by plugging the positive side of the battery in the B part, we lowered the electrostatic energy of its electrons by $-|q| V_0$ with respect to the electrostatic energy in system A, where $|q|$ is the modulus of the electron's charge (the elementary charge). The higher chemical potential of system B was thus balanced by its lower electrostatic potential, that is:

$$\mu_0 = |q| V_0 \qquad (1.41)$$

Thus, the net result is that the **total** potential is the same in both systems. Importantly, the net current is zero, because the drift current from B to A due to the battery is exactly cancelled by the diffusion current from A to B due to the difference in the chemical potentials.

The situation of Figure 1.6c is similar to the situation of Figure 1.5b, with the battery playing the role of the gravitational potential. There is, however, an important difference between elements without charge, as in Figure 1.5, and elements with charge, as in Figure 1.6.

To appreciate this difference, go back to the initial situation of Figure 1.6a, but now do not freeze the experiment: let the charges flow until equilibrium is reached. At the beginning of the process, there is only a difference in chemical potential (but no difference in electrostatic potential), so the situation in the beginning is similar to Figure 1.5a when the gases have different chemical potentials (and there is no difference in gravitational potential). So, one might think that equilibrium in Figure 1.6a is reached in the same way as the equilibrium in Figure 1.5a, that is, when the chemical potentials equalize. But that is not the case: because the elements have charge, the diffusion current causes a charge imbalance between the systems, and this charge imbalance create an electric field and, hence, an electrostatic potential difference. Therefore, contrary to the equilibrium condition of Figure 1.5a, the chemical potentials are never equalized in Figure 1.6a. Let me detail this crucial process a bit more.

The equilibrium condition reached when the charges of Figure 1.6a are allowed to diffuse is shown in Figure 1.6d. In the beginning, the systems of Figure 1.6a are neutral. As charges flow from system B to system A, the former will be become positively charged, whereas the latter will become negatively charged. As a matter of fact, due to a process that will be discussed in Chapter 2, the charges accumulate at the interface between the systems, creating a kind of internal capacitor. These charges set up an internal electric field, in the same way that the charges of a capacitor set up an electric field.

This internal electric field thus induces a drift current, at the opposite direction of the diffusion current (see Figure 1.6d). Equilibrium is reached when the two currents cancel each other.

We know that there is an electrostatic potential difference due to the field in a capacitor. Likewise, there will be an electrostatic potential difference due to the internal field. As expected, this potential difference is the good old $-V_0$ set up by the battery in Figure 1.6b and Figure 1.6c. As a matter of fact, the situation of Figure 1.6d is similar to the situation of Figure 1.6c; the only difference is that there is no external battery in Figure 1.6d: the electrostatic potential difference is due solely to the internal field. And if you plug a voltmeter across the systems, the reading will be zero, because the total potential (chemical + electrostatic) is the same in both systems, though neither the chemical, nor the electrostatic potentials are individually the same in both systems. Finally, notice that the system, as a whole, is not charged, because there is only an internal flow of charge (every negative charge in system A is compensated by a positive one in system B, as this negative charge in A is due to an electron flowing from B to A). The physical situation illustrated in Figure 1.6d is akin to the equilibrium condition in a **p-n** junction, as will be seen in Chapter 3.

We have reached the conclusion that a difference in chemical potential of charged elements results in the creation of an internal electrostatic potential. This means that, for particles carrying charge, the chemical and electrostatic potentials always come together. Thus, it does not make too much sense to talk about the chemical potential on its own, or the electrostatic potential on its own, since it is the total potential that governs the transport of charges. That this is indeed the case is conspicuous in Figure 1.6d: there is both a chemical potential difference and an electrostatic potential difference, but a voltmeter will give a zero potential difference, because the voltmeter measures the total potential, which is the same in both systems.

This total potential is named the "electrochemical potential". As we have just seen, equilibrium is reached when the electrochemical potential is equalized across the systems. The electrochemical potential of electrons is given by:

$$\mu_{eq} = \mu - | q | V \tag{1.42}$$

where μ is the chemical potential, V is the electrostatic potential in units of Volts, and the negative sign is due to the negative charge of electrons. Just for the sake of argument, if gravity also played a role, one would need to add it to the total potential:

$$\mu_{total} = \mu - |q| V + mgh \tag{1.43}$$

But, for our purposes, gravity does not play a role, so it is the electrochemical potential by itself that matters (that is not to say that gravity never plays a role; for example, it plays an important role in osmosis).

Unfortunately, the literature on semiconductors is not very consistent with notation, and often we find the same symbol being ascribed to the chemical and electrochemical potentials. Moreover, some books use the term "internal potential" instead of "electrochemical potential", so one needs to be careful to identify what potential is being described.

To simplify the notation, up to the end of this chapter, I will use the symbol μ for the electrochemical potential and use other symbols for the chemical potential when necessary. In the next chapter we will need to change the symbols again.

Box 5   The electron volt

In the context of semiconductor physics, energies are almost always quoted in units of electron volts (*e. V.*). By definition 1 *e. V.* $= |q|$ J, where $|q|$ is the numerical value of the elementary charge ($|q| = 1.602176634 \times 10^{-19}$) and J is joules.

Suppose you are told that an electron loses 1 *e. V.* when it crosses an element (e.g. a resistor). What is the voltage associated with this energy difference? Recall that voltage is energy (in units of joules) per unit of coulombs. So, the voltage is the energy of the electron in joules divided by the elementary charge, that is:

$$V = \frac{E[J]}{|q|[C]}$$

where $E[J]$ is the energy in units of joules and $|q|[C]$ is the fundamental charge in units of coulombs. But if an electron loses 1 *e. V*, then the voltage drop is:

$$V = \frac{E[J]}{|q|[C]} = \frac{1 \times |q|}{|q|} \left[\frac{J}{C}\right] = 1 \left[\frac{J}{C}\right].$$

That is the beauty of the electron volt: if you know the energy in electron volts, then the conversion to volts involves only a change of units, but the numerical value remains the same.

1.5   The partition function – part II

We have seen that systems in diffusive contact are described in terms of two properties: energy U and number of particles N. Thus, we need to include the new property in the derivation of the partition function. The logic is about the same as in section 1.3: we start by considering a system A that is in thermal and diffusive contact with a reservoir, and then express the ratio of the probability of finding system A in a given state β to the probability of finding system A in a given state α. The energy and number of elements of state β will be respectively denoted by ε_β and n_β, whereas the energy and number of elements of state α will be respectively denoted by ε_α and n_α. As in section 1.3, the ratio of probabilities is the ratio between the multiplicities of the reservoir at these two conditions. Thus:

$$\frac{P(\beta)}{P(\alpha)} = \frac{g_R(U_0 - \varepsilon_\beta, N_0 - n_\beta)}{g_R(U_0 - \varepsilon_\alpha, N_0 - n_\alpha)} \tag{1.44}$$

Remember that β and α are states of system A, but that g_R is the multiplicity of the reservoir. The energy and number of elements of the total system (system A + reservoir) are U_0 and N_0, respectively.

The procedure to find the partition function and the absolute probability $P(c)$ of a given state c is analogous to the procedure shown in section 1.3. We begin by expressing Equation (1.44) in terms of the entropy:

$$\frac{P(\beta)}{P(\alpha)} = \frac{\exp\left[\dfrac{S_R(U_0 - \varepsilon_\beta, N_0 - n_\beta)}{k_B}\right]}{\exp\left[\dfrac{S_R(U_0 - \varepsilon_\alpha, N_0 - n_\alpha)}{k_B}\right]}$$

$$= \exp\left[\frac{S_R(U_0 - \varepsilon_\beta, N_0 - n_\beta) - S_R(U_0 - \varepsilon_\alpha, N_0 - n_\alpha)}{k_B}\right] \tag{1.45}$$

and, again, expand the entropy in a Taylor series, retaining only the first order terms. The difference is that, since the entropy is now a function of two variables, there are two first order terms:

$$S_R(U_0 - \varepsilon, N_0 - n) \cong S_R(U_0, N_0) - \varepsilon \frac{\partial S_R}{\partial U} - n \frac{\partial S_R}{\partial N} \tag{1.46}$$

Using Equation (1.23) and Equation (1.37), Equation (1.46) reads:

$$S_R(U_0 - \varepsilon, N_0 - n) = S_R(U_0, N_0) - \frac{\varepsilon}{T} + \mu \frac{n}{T} \tag{1.47}$$

Plugging it into Equation (1.45):

$$\frac{P(\beta)}{P(\alpha)} = \frac{\exp\left[\dfrac{-\left(\varepsilon_\beta - n_\beta \mu\right)}{k_B T}\right]}{\exp\left[\dfrac{-\left(\varepsilon_\alpha - n_\alpha \mu\right)}{k_B T}\right]} \tag{1.48}$$

Once again, to find the absolute value of the probability, we need to find the partition function, which is the sum of the exponentials over all states of system A. Thus:

$$\zeta(T,\mu) = \sum_c \exp\left[\frac{-\left(\varepsilon_c - n_c \mu\right)}{k_B T}\right] \tag{1.49}$$

where $\zeta(T,\mu)$ is the partition function of system A in thermal and diffusive equilibrium with a reservoir.

Notice that I used another symbol for the partition function: instead of Z (when there was only thermal equilibrium), I now used ζ (when there is both thermal and diffusive equilibrium). This change is necessary not only to differentiate between the two cases, but also because whereas Z is a function of only T, ζ is a function of T and μ.

Even though $\zeta(T,\mu)$ is just the partition function of a system in both thermal and diffusive equilibrium with a reservoir, it is usually given a different name. Actually, it goes by two names: either "grand sum", or "Gibbs sum". I shall adopt the latter.

Finally, using the Gibbs sum, the absolute probability of finding system A in a given state c, when system A is in thermal and diffusive equilibrium with a reservoir, is given by:

$$P(c) = \frac{\exp\left[\dfrac{-\left(\varepsilon_c - n_c \mu\right)}{k_B T}\right]}{\zeta(T,\mu)} \tag{1.50}$$

Thus, the probability of system A to be found in a state c depends on the properties of state c (which are its energy ε_c and number of elements n_c), on the temperature T and potential μ of the reservoir. Finally, notice that information about system A (is it a solid, a gas? which solid, which gas?) enters in the Gibbs sum, since it depends on all states of system A, and their respective properties.

1.6   Example of application: energy and number of elements of a system

Equation (1.31) and Equation (1.50) are two important results that we have obtained so far. They give the probability of a system to be in a given state, when this system is either only in thermal equilibrium with a reservoir (Equation (1.31)) or in both thermal and diffusive equilibrium with a reservoir (Equation (1.50)). But what is effectively the state of the system? The answer given by statistical physics is a resounding: I don't know!! We have only the probabilities. But often this is all we need to deduce important properties of macroscopic systems, and in this section I will show you a paradigmatic example of application of these concepts.

To make the logic clear, we will consider a system in only thermal, but not diffusive, equilibrium with a reservoir, so we will use Equation (1.31) instead of Equation (1.50). Thus, our system example consists of a single particle trapped in a box (the box is closed, so no diffusive contact). The box is in thermal equilibrium with the reservoir at temperature T. Suppose that your job is to determine the state of your particle. For the sake of argument, we will assume that you can look at the particle and determine its state without affecting it. So, suppose that you look at your particle and find it in state a. Then you wait a bit and look at it again. Since the particle is interacting with the reservoir, it is more likely than not that you will find it in another state, say state b, when you look at it again.

Ok, so you do this measurement several times, and write down the state you found it in. You could do something important with this data: you could, for example, find the mean value of the energy (remember that if you know the state you also know the energy). To do that, you just take the average of all the energy values you wrote down. We will denote the average of the energy by $\langle U \rangle$. How can we express this thought experiment in mathematical terms? Suppose that you took M measurements in total, and that the state c, whose energy is ε_c, appeared M_c times in your measurement. Thus, adding up the contributions of all the states that appeared in your measurements, the average is:

$$\langle U \rangle = \frac{\sum_c \varepsilon_c M_c}{M}$$

where $\sum_c M_c = M$.

Say your life is not that very interesting, and you are OK with the prospect of having to do these measurements billions and billions of times. If you do it billions and billions of times, then the ratio M_c/M will approximate the probability $P(c)$ of finding the state c (remember that M_c is the number of times that state c showed up). Thus, we can express $\langle U \rangle$ in terms of the probabilities:

$$\langle U \rangle = \sum_c P(c)\varepsilon_c \qquad (1.51)$$

But since $P(c)$ is the probability of finding a given state c is a system in thermal equilibrium with a reservoir, it is given by our good old Equation (1.31). Thus, substituting Equation (1.31) into Equation (1.51):

$$\langle U \rangle = \sum_c \varepsilon_c P(c) = \frac{\sum_c \varepsilon_c \exp\left[-\frac{\varepsilon_c}{k_B T}\right]}{Z(T)} \qquad (1.52)$$

The job of statistical physics kind of ended here. Now we need information from other areas of physics to be able to compute the sum in Equation (1.52), because we need to know what the states of system A are, and what their energies are. Fortunately, we picked an example for which we already know how to compute the states and energies. Indeed, we have already found the states of a particle trapped in a box: the states are represented by natural numbers and their corresponding energies are given by Equation (1.8). There is a difference though: Equation (1.8) was derived by considering that the particle was trapped in a one-dimensional box, but a true box is a three-dimensional box. To find the expression for a three-dimensional box, we can follow the same logic that led to Equation (1.8), but now we must require that the phases accumulated in round trips in all three dimensions are multiples of 2π. The three-dimensional wave function is:

$$\psi(x, y, z) = \exp\left[i\left(k_x x + k_y y + k_z z\right)\right] \qquad (1.53)$$

Now we require that the phase accumulation in all three directions is a multiple of 2π, which leads to three independent requirements, one for each component of the wavevector:

$$k_x = \frac{c_x \pi}{L_x}, k_y = \frac{c_y \pi}{L_y}, k_z = \frac{c_z \pi}{L_z} \qquad (1.54)$$

where c_x, c_y, and c_z are natural numbers, and L_x, L_y and L_z are the dimensions of the box. To simplify the equations, let us assume that the box is a cube, thus: $L_x = L_y = L_z = L$, and:

$$k_x = \frac{c_x \pi}{L}, k_y = \frac{c_y \pi}{L}, k_z = \frac{c_z \pi}{L} \qquad (1.55)$$

As discussed in section 1.1, the mechanical momentum is proportional to the wavevector. Thus:

$$\vec{p} = \frac{hk_x}{2\pi}\hat{x} + \frac{hk_y}{2\pi}\hat{y} + \frac{hk_z}{2\pi}\hat{z} \tag{1.56}$$

where ^ denotes a unit vector. From Newtonian mechanics, we know that the kinetic energy is:

$$E = \frac{|\vec{p}|^2}{2m} = \frac{\left(\frac{hk_x}{2\pi}\right)^2 + \left(\frac{hk_y}{2\pi}\right)^2 + \left(\frac{hk_z}{2\pi}\right)^2}{2m} \tag{1.57}$$

where m is again the mass of the particle

Thus, with the help of Equation (1.55), the allowed energies of a particle trapped in a box are given by:

$$E_{c_x,c_y,c_z} = \frac{h^2}{8L^2 m}\left(c_x^2 + c_y^2 + c_z^2\right) \tag{1.58}$$

Equation (1.58) is the three-dimensional version of Equation (1.8). In the 3D system, each combination of c_x, c_y and c_z represents a different state, so we denote each state by a triplet of numbers (c_x, c_y, c_z). For example, the ground state, that is, the state with the lowest energy, is the state $(1, 1, 1)$. Another important difference between 1D and 3D systems is that, while all states in the 1D system had a different energy, two or more states in a 3D system can have the same energy. For example, the states $(1, 1, 2)$, $(1, 2, 1)$ and $(2, 1, 1)$ have the same energy $E = \frac{h^2}{8L^2 m}6$.

Now that we know the states and their energies, we can compute the partition function:

$$Z(T) = \sum_c \exp\left[-\frac{\varepsilon_c}{k_B T}\right] = \sum_{c_x=1}^{\infty}\sum_{c_y=1}^{\infty}\sum_{c_z=1}^{\infty} \exp\left[-\frac{\frac{h^2}{8L^2 m}\left(c_x^2 + c_y^2 + c_z^2\right)}{k_B T}\right] \tag{1.59}$$

Notice that the sum involves all possible states, that is, all possible combinations (c_x, c_y, c_z) of natural numbers. In section 2.3 I will show you a trick to compute this sum, but here I am more interested in showing you how the different data are put together. So, to avoid distraction, for now I will just quote the result of the sum, and defer the proof to section 2.3. The result is:

$$Z(T) = \sum_{c_x = 1}^{\infty} \sum_{c_y = 1}^{\infty} \sum_{c_z = 1}^{\infty} \exp\left[-\frac{\frac{h^2}{8L^2 m}\left(c_x^2 + c_y^2 + c_z^2\right)}{k_B T}\right] = \frac{L^3}{\left(\frac{h^2}{2\pi m k_B T}\right)^{3/2}}$$

$$(1.60)$$

Notice that L^3 is the volume of the box. Thus, the partition function of the system "one particle in a box" depends on the volume of the box, the mass of the particle, and the temperature.

Now we have an explicit equation for the average energy:

$$\langle U \rangle = \frac{\sum_{c_x = 1}^{\infty}\sum_{c_y = 1}^{\infty}\sum_{c_z = 1}^{\infty} \frac{h^2}{8L^2 m}\left(c_x^2 + c_y^2 + c_z^2\right) \exp\left[-\frac{\frac{h^2}{8L^2 m}\left(c_x^2 + c_y^2 + c_z^2\right)}{k_B T}\right]}{\frac{L^3}{\left(\frac{h^2}{2\pi m k_B T}\right)^{3/2}}}$$

$$(1.61)$$

Computing the sum, we find:

$$\langle U \rangle = \frac{3}{2} k_B T \qquad (1.62)$$

This is a famous result: you probably have already heard that the average energy per atom of an ideal gas is $\frac{3}{2} k_B T$. An ideal gas is a gas where the particles do not interact with each other. So, if the particles do not interact, instead of measuring the energy of a single particle lots of times, like we did in our thought experiment, you could get several identical particles and measure their energies at the same time, and the result would be the same. Thus, if you have N particles in this gas, Equation (1.62) is telling you that the energy of the gas is $\frac{3}{2} k_B T N$.

We can straightforwardly generalize the logic exemplified here to include diffusive contact: all we need to do is to replace the partition function by the Gibbs sum. If there is diffusive contact, we could, for example, compute the average number of elements:

$$\langle N \rangle = \sum_c n_c P(c) = \sum_c n_c \frac{\exp\left[\frac{-(\varepsilon_c - n_c \mu)}{k_B T}\right]}{\zeta(T, \mu)} \qquad (1.63)$$

Likewise, the average energy if there is diffusive equilibrium is:

$$\langle U \rangle = \sum_c \varepsilon_c P(c) = \sum_c \varepsilon_c \frac{\exp\left[\dfrac{-(\varepsilon_c - n_c \mu)}{k_B T}\right]}{\zeta(T, \mu)} \tag{1.64}$$

To conclude this section, let us summarize the nomenclature we have been using, and connect the fictitious system of coins to real systems made of electrons (a more detailed connection between both systems is given in section 1.8). It is important that you do not confuse the state of an electron with the state of the system. In the example of the coins, one coin plays the role of an electron, and the state of the coin is given by *heads* or *tails*. If there are more than one coin, then the state of the system is given by the combinations of states of all coins. For example, in a two-coins system, if one coin is *heads* and the other is *tails*, then the state of the *system* is HT. Analogously, in a system with a single electron trapped in a box, the state of the system coincides with the state of the electron, and both are given by the triplet (c_x, c_y, c_z). But, if there are more than one electron in the system, then the state of the system is given by combinations of states of the electrons. For example, in a system with two electrons, if one electron is in the state $(1, 1, 1)$ and the other is in the state $(1, 1, 2)$, then the state of the system is given by the state "one electron in state $(1, 1, 1)$ and another in state $(1, 1, 2)$". In the next section, we will define a convenient notation to denote states of systems with more than one electron. An important feature to notice is that, because electrons are indistinguishable, it does not matter which one is in state $(1, 1, 1)$ and which one is in state $(1, 1, 2)$, which means that there is only one state of the total system where "one electron is in state $(1, 1, 1)$ and another is in state $(1, 1, 2)$". The effect of the indistinguishability of electrons plays an important role in the Fermi–Dirac distribution, which is a key concept of statistical physics pertinent to the field of semiconductors, and that is the subject of the next section.

1.7 The Fermi–Dirac distribution

Now we turn our attention to a major result of statistical physics. Our goal is to determine the probability of an electron to be in a given state of interest, when the system has a large number (more specifically, a macroscopic number) of electrons. Our system of interest is again the system "electrons trapped in a box", in thermal equilibrium with a reservoir. In principle, we could use the equation involving the partition function (Equation (1.31)) to find the probability of an electron to be in a given state of interest.

But recall that the probability of Equation (1.31) refers to the state of the system, which is the combination of states of all electrons. So, to use Equation (1.31), we would first have to identify all the states of the system that has one electron in the state of interest, use Equation (1.31) to work out the probability of these states of the system, and then add up all these probabilities of states of the system where there is an electron in the state of interest. Though in principle possible, this procedure becomes cumbersome when there is a macroscopic number of electrons. So, we need another strategy to find the probability of an electron to be in a given state of interest.

To define the strategy for this calculation, let us begin by gaining insight into how the multiplicity function of a system "lots of electrons inside a box" is built up. Though we are ultimately interested in systems with a macroscopic number of electrons, to begin with, we are going to consider small systems, so as to get an intuition about how the multiplicity is built up, and later we generalize to macroscopic systems. Furthermore, and again for the sake of simplicity, I will ignore the spin (ignoring the spin does not alter the main result we want to derive in this section).

First, recall that the multiplicity g is a function of two variables: the energy U and the number of particles N, that is: $g(U, N)$. We also know what the states of electrons in the system are: they are indexed by three natural numbers c_x, c_y, c_z. And we also know that the energy of each state is given by:

$$U_{c_x, c_y, c_z} = \frac{h^2}{8L^2 m}\left(c_x^2 + c_y^2 + c_z^2\right)$$

To get an intuition about how to construct the multiplicity, let us evaluate it for a few examples (remember that we are ignoring the spin, but if you want very much to know what the difference in the reasoning would be if we had to include the spin, you will notice after the first example that the effect of the spin is kind of trivial here: it just doubles the number of states). Thus, what is the value of $g(U_{1, 1, 1}, 1)$? Recall that the multiplicity is the number of states, so the value of $g(U_{1, 1, 1}, 1)$ is the number of states of the system whose energy is equal to $\frac{3h^2}{8L^2 m}$ and number of particles equal to 1. How many states satisfy these two conditions? Well, if there is only one particle, that is, just one electron, then obviously there is only one state of the system that satisfies these conditions: the state where there is a single electron, and it is in the state $c_x = 1$, $c_y = 1$, $c_z = 1$. So, our answer is $g(U_{1, 1, 1}, 1) = 1$ (if you considered the spin, then the answer would be 2). To make it a bit more organized, let us express this result listing the possible states explicitly, like this:

$$g(U_{1,1,1}, 1) = 1$$

Possible states :

$$(1, 1, 1)$$

Now, let us evaluate $g(U_{1, 1, 2}, 1)$. We still have only one electron, but now there are three states having the energy $U_{1,1,2} = \frac{6h^2}{8L^2 m}$, so $g(U_{1, 1, 2}, 1) = 3$. Listing the states explicitly:

$$g(U_{1,1,2}, 1) = 3$$

Possible states :

$$(1, 1, 2)$$
$$(1, 2, 1)$$
$$(2, 1, 1)$$

Let us complicate a bit more now: let us compute a case with two electrons. For example, we can deduce straight away that $g(U_{1, 1, 1}, 2) = 0$ because there is no possible state of the system where we have two electrons but the total energy is $U_{1, 1, 1}$. To see this, remember that if there is one electron in the state $(1, 1, 1)$, then the energy of the other electron already precludes the possibility of the system having the energy $U_{1, 1, 1}$ (since the other electron must have an energy different than zero). What about $g(2 U_{1, 1, 1}, 2)$? Well, if you take spin into account, then there are two ground states $(1, 1, 1)$, one for each spin, so there is one state of the total system where both electrons are in these two ground states. Since the energy of each electron is $U_{1, 1, 1}$, this state of the system has energy $2 U_{1, 1, 1}$. So, the "official" answer is $g(2 U_{1, 1, 1}, 2) = 1$. But, since we are ignoring the spin, in our case the second electron must be in another state due to Pauli exclusion principle, which precludes the possibility of having the total energy equal to $2 U_{1, 1, 1}$. Thus, we still have $g(2 U_{1, 1, 1}, 2) = 0$ in our system that ignores the spin.

Let me give one more example, still with two electrons, but now with a total energy $U = U_{1, 1, 1} + U_{1, 1, 2}$. In this case, we need one electron in the ground state $(1, 1, 1)$ and the other in one of the three next excited states. So, there is a total possibility of three states satisfying the requirement that total energy is $U = U_{1, 1, 1} + U_{1, 1, 2}$ and $N = 2$, so:

$$g(U_{1,1,1} + U_{1,1,2}, 2) = 3$$

Possible states :

$$(1, 1, 1; 1, 1, 2)$$
$$(1, 1, 1; 1, 2, 1)$$
$$(1, 1, 1; 2, 1, 1)$$

where the state $(A; B)$ reads "one electron in the state A and another electron in the state B". For example, the state $(1, 1, 1; 1, 1, 2)$ is the state "one electron in the state 1, 1, 1 and the other is in the state 1, 1, 2". Notice that the state $(A; B)$ is identical to state $(B; A)$, which is a consequence of the indistinguishability of electrons. For example, the list above has the state $(1, 1, 1; 1, 1, 2)$, but not the state $(1, 1, 2; 1, 1, 1)$. If we had added both the state $(1, 1, 1; 1, 1, 2)$ and the state $(1, 1, 2; 1, 1, 1)$, we would be counting the same state twice.

These examples are useful to emphasize once more a subtlety in the nomenclature: when we say state, we may mean two different things: we may mean the state of a single electron, given by the three natural numbers, or we may mean state of the system, which is a combination of the states of electrons, and that we are denoting by listing the electron states, as in $(1, 1, 1; 1, 1, 2)$ in the list above. This ambiguity may lead to confusion; so, to remove it, from now on, I will use the term "orbital" to denote the state of a single electron, and the term "state" to denote the state of the system. Thus, we can say that $(A; B)$ is the state (of the system) "one electron is in the orbital A and the other is in the orbital B". Furthermore, we can re-express our goal in this section using this new terminology: instead of saying that we want to find the probability of an electron to be in a state of interest, we will say that we want to find the probability of an electron to be in an orbital of interest.

To make sure we get a good intuition about these lists of possible states, let us give one more example, now for $U = 2U_{1,\ 1,\ 2}$, $N = 2$. In this case, we have:

$$g\left(2U_{1,1,2}, 2\right) = 6$$

$$Possible\ states:$$

$$(1, 1, 2; 1, 2, 1)$$

$$(1, 1, 2; 2, 1, 1)$$

$$(1, 2, 1; 2, 1, 1)$$

$$(1, 1, 1; 1, 2, 2)$$

$$(1, 1, 1; 2, 1, 2)$$

$$(1, 1, 1; 2, 2, 1)$$

Check for yourself that all states above have the energy $U = 2U_{1,\ 1,\ 2}$. These examples illustrate how we build the lists of possible states, and hence find the multiplicity, for a given pair of energy and number of electrons.

Now that we have a good idea about how to draw up these lists, let us change the notation a bit, to make it more convenient for systems with a large number of electrons. When we had two electrons, we expressed the state by listing the orbitals separated by a semicolon, inside a parenthesis. If we had three electrons, we would have three orbitals, separated by semicolons, and so on. Instead of doing that, let us use a super-parenthesis and fill it with all possible orbitals, like this:

$$(1,1,1;1,1,2;1,2,1;2,1,1;1,2,2;2,1,2;2,2,1;1,1,3;\ \ldots)$$

where the ... indicates that the list goes on and on, to infinity.

With this notation, we can indicate the state by denoting which orbitals are occupied. For example, the state $(1,1,1;1,1,2)$ is the state with orbitals $1,1,1$ and $1,1,2$ occupied by electrons, with all the others unoccupied. Thus, we can denote this state by putting the letter o, for "occupied", on top of orbitals $1,1,1$ and $1,1,2$ and the letter u, for "unoccupied", on top of the other orbitals, like this:

$$(1,1,1;1,1,2)$$

$$\leftrightarrow \begin{pmatrix} o & o & u & u & u & u & u & u & u \\ 1,1,1 & 1,1,2 & 1,2,1 & 2,1,1 & 1,2,2 & 2,1,2 & 2,2,1 & 1,1,3 & \ldots \end{pmatrix}$$

As another example, using this notation, the list of all possible states for $U = 2\,U_{1,\,1,\,2}$, $N = 2$ reads:

$$g\big(2\,U_{1,1,2}, 2\big) = 6$$

Possible states :

$$\begin{pmatrix} u & o & o & u & u & u & u & u & u \\ 1,1,1 & 1,1,2 & 1,2,1 & 2,1,1 & 1,2,2 & 2,1,2 & 2,2,1 & 1,1,3 & \ldots \end{pmatrix}$$

$$\begin{pmatrix} u & o & u & o & u & u & u & u & u \\ 1,1,1 & 1,1,2 & 1,2,1 & 2,1,1 & 1,2,2 & 2,1,2 & 2,2,1 & 1,1,3 & \ldots \end{pmatrix}$$

$$\begin{pmatrix} u & u & o & o & u & u & u & u & u \\ 1,1,1 & 1,1,2 & 1,2,1 & 2,1,1 & 1,2,2 & 2,1,2 & 2,2,1 & 1,1,3 & \ldots \end{pmatrix}$$

$$\begin{pmatrix} o & u & u & u & o & u & u & u & u \\ 1,1,1 & 1,1,2 & 1,2,1 & 2,1,1 & 1,2,2 & 2,1,2 & 2,2,1 & 1,1,3 & \ldots \end{pmatrix}$$

$$\begin{pmatrix} o & u & u & u & u & o & u & u & u \\ 1,1,1 & 1,1,2 & 1,2,1 & 2,1,1 & 1,2,2 & 2,1,2 & 2,2,1 & 1,1,3 & \ldots \end{pmatrix}$$

$$\begin{pmatrix} o & u & u & u & u & u & o & u & u \\ 1,1,1 & 1,1,2 & 1,2,1 & 2,1,1 & 1,2,2 & 2,1,2 & 2,2,1 & 1,1,3 & \dots \end{pmatrix}$$

Having sorted the notation, let us move on to actual physics. Suppose that the system "box", with energy $U = 2U_{1,\,1,\,2}$ and number of electrons $N = 2$, is our total system. Our goal is to obtain the probability of finding a given orbital occupied. As an example, let us use the orbital 2, 1, 1. Furthermore, again for the sake of simplicity, let us suppose for now that the system is isolated from the universe.

In a case with only a handful of electrons, like the ones we have been considering, it is quite trivial to find this probability: just count how many states have the orbital of interest occupied and divide it by the number of all possible states. In our example, the list above has a total of six states, out of which two have the orbital 2, 1, 1 occupied (the second and third states in the list above). Since, according to the fundamental assumption of statistical physics, all states have equal probability, we conclude that, in our system example, the probability of finding the orbital 2, 1, 1 occupied is 2/6.

Now, bear with me a little bit, and let me do some rearrangement in the list above. This may look a bit silly now, but it will be helpful when we consider large systems. First, I will move the orbital of interest, the orbital 2, 1, 1, to the first place of the list in the parentheses, like this:

$$g\big(2U_{1,1,2}, 2\big) = 6$$

Possible states :

$$\begin{pmatrix} u & u & o & o & u & u & u & u & u \\ 2,1,1 & 1,1,1 & 1,1,2 & 1,2,1 & 1,2,2 & 2,1,2 & 2,2,1 & 1,1,3 & \dots \end{pmatrix}$$

$$\begin{pmatrix} o & u & o & u & u & u & u & u & u \\ 2,1,1 & 1,1,1 & 1,1,2 & 1,2,1 & 1,2,2 & 2,1,2 & 2,2,1 & 1,1,3 & \dots \end{pmatrix}$$

$$\begin{pmatrix} o & u & u & o & u & u & u & u & u \\ 2,1,1 & 1,1,1 & 1,1,2 & 1,2,1 & 1,2,2 & 2,1,2 & 2,2,1 & 1,1,3 & \dots \end{pmatrix}$$

$$\begin{pmatrix} u & o & u & u & o & u & u & u & u \\ 2,1,1 & 1,1,1 & 1,1,2 & 1,2,1 & 1,2,2 & 2,1,2 & 2,2,1 & 1,1,3 & \dots \end{pmatrix}$$

$$\begin{pmatrix} u & o & u & u & u & o & u & u & u \\ 2,1,1 & 1,1,1 & 1,1,2 & 1,2,1 & 1,2,2 & 2,1,2 & 2,2,1 & 1,1,3 & \dots \end{pmatrix}$$

$$\begin{pmatrix} u & o & u & u & u & u & o & u & u \\ 2,1,1 & 1,1,1 & 1,1,2 & 1,2,1 & 1,2,2 & 2,1,2 & 2,2,1 & 1,1,3 & \dots \end{pmatrix}$$

So, the list is still the same, I just moved the orbital of interest to the first place. I will do one further change, and move all states that have the orbital of interest occupied to the top of the list, like this:

$$g\big(2\,U_{1,1,2}, 2\big) = 6$$

Possible states :

$$\begin{pmatrix} o & u & o & u & u & u & u & u & u \\ 2,1,1 & 1,1,1 & 1,1,2 & 1,2,1 & 1,2,2 & 2,1,2 & 2,2,1 & 1,1,3 & \dots \end{pmatrix}$$

$$\begin{pmatrix} o & u & u & o & u & u & u & u & u \\ 2,1,1 & 1,1,1 & 1,1,2 & 1,2,1 & 1,2,2 & 2,1,2 & 2,2,1 & 1,1,3 & \dots \end{pmatrix}$$

$$\begin{pmatrix} u & u & o & o & u & u & u & u & u \\ 2,1,1 & 1,1,1 & 1,1,2 & 1,2,1 & 1,2,2 & 2,1,2 & 2,2,1 & 1,1,3 & \dots \end{pmatrix}$$

$$\begin{pmatrix} u & o & u & u & o & u & u & u & u \\ 2,1,1 & 1,1,1 & 1,1,2 & 1,2,1 & 1,2,2 & 2,1,2 & 2,2,1 & 1,1,3 & \dots \end{pmatrix}$$

$$\begin{pmatrix} u & o & u & u & u & o & u & u & u \\ 2,1,1 & 1,1,1 & 1,1,2 & 1,2,1 & 1,2,2 & 2,1,2 & 2,2,1 & 1,1,3 & \dots \end{pmatrix}$$

$$\begin{pmatrix} u & o & u & u & u & u & o & u & u \\ 2,1,1 & 1,1,1 & 1,1,2 & 1,2,1 & 1,2,2 & 2,1,2 & 2,2,1 & 1,1,3 & \dots \end{pmatrix}$$

We still have the same list of possible states. Now, still considering the same list, let us make an imaginary division of the system. You may consider the problem through the following perspective: the system "box with electrons" is essentially the group of all possible orbitals of the box. Without changing anything physically, I can describe my system "box" as being composed of two subsystems: one containing only the orbital of interest, and the other containing all the other orbitals. Notice that this is merely a conceptual division, not a physical one. We can represent this conceptual division by highlighting the two subsystems:

$$g(2U_{1,1,2}, 2) = 6$$

Possible states:

$$
\begin{aligned}
&\begin{pmatrix} \overset{o}{2,1,1} & \overset{u}{1,1,1} & \overset{o}{1,1,2} & \overset{u}{1,2,1} & \overset{u}{1,2,2} & \overset{u}{2,1,2} & \overset{u}{2,2,1} & \overset{u}{1,1,3} & \overset{u}{...} \end{pmatrix}\\[4pt]
&\begin{pmatrix} \overset{o}{2,1,1} & \overset{u}{1,1,1} & \overset{u}{1,1,2} & \overset{o}{1,2,1} & \overset{u}{1,2,2} & \overset{u}{2,1,2} & \overset{u}{2,2,1} & \overset{u}{1,1,3} & \overset{u}{...} \end{pmatrix}\\[4pt]
&\begin{pmatrix} \overset{u}{2,1,1} & \overset{u}{1,1,1} & \overset{o}{1,1,2} & \overset{o}{1,2,1} & \overset{u}{1,2,2} & \overset{u}{2,1,2} & \overset{u}{2,2,1} & \overset{u}{1,1,3} & \overset{u}{...} \end{pmatrix}\\[4pt]
&\begin{pmatrix} \overset{u}{2,1,1} & \overset{o}{1,1,1} & \overset{u}{1,1,2} & \overset{u}{1,2,1} & \overset{o}{1,2,2} & \overset{u}{2,1,2} & \overset{u}{2,2,1} & \overset{u}{1,1,3} & \overset{u}{...} \end{pmatrix}\\[4pt]
&\begin{pmatrix} \overset{u}{2,1,1} & \overset{o}{1,1,1} & \overset{u}{1,1,2} & \overset{u}{1,2,1} & \overset{u}{1,2,2} & \overset{o}{2,1,2} & \overset{u}{2,2,1} & \overset{u}{1,1,3} & \overset{u}{...} \end{pmatrix}\\[4pt]
&\begin{pmatrix} \overset{u}{2,1,1} & \overset{o}{1,1,1} & \overset{u}{1,1,2} & \overset{u}{1,2,1} & \overset{u}{1,2,2} & \overset{u}{2,1,2} & \overset{o}{2,2,1} & \overset{u}{1,1,3} & \overset{u}{...} \end{pmatrix}
\end{aligned}
$$

SYSTEM A · SYSTEM B

where I labelled the system "orbital of interest" as system A, and the other as system B.

Now comes the conceptual trick that will allow us to easily obtain the probability of finding a given orbital occupied in a large system. I will again use the small system to show the trick, as it is easier to grasp it in small systems. The trick is to express the probability of finding the orbital of interest occupied in terms of properties of the system comprising the other orbitals, which we labelled system B. This can be done by noticing that the properties of the two systems are complementary. In our example, the energy of the total system is $U = 2U_{1,1,2}$ and the number of electrons is $N = 2$. That means that, if system A has the orbital of interest 2, 1, 1 occupied, then necessarily system B can have only one orbital occupied, since the total number of electrons in our example is $N = 2$. On the other hand, if the orbital of interest is unoccupied, then system B necessarily has two orbitals occupied. Likewise, the total energy, which in our example is $U = 2U_{1,1,2}$, also adds a constraint. Indeed, if the orbital of interest is occupied, that means that the energy of system A is $U_{1,\,1,\,2}$ (recall that $U_{2,\,1,\,1} = U_{1,\,1,\,2}$). Thus, the energy of system B must also be $U_{1,\,1,\,2}$. If, however, the orbital of interest is not occupied, then the energy of system A is zero, which means that the energy of system B must be $2\,U_{1,\,1,\,2}$.

Bearing in mind these considerations, it becomes clear that the question "how many states of the total system have the orbital of interest occupied?" is logically equivalent to the question "how many states of system B have the complementary properties?" As we will see shortly, it is much easier to answer the latter question.

To solidify the idea, let us show the list of states of system B with the complementary properties of our example. Recall that, in our example, when the orbital of interest is occupied, then the complementary properties of system B are $U_B = 2\,U_{1,\,1,\,2} - U_{2,\,1,\,1} = U_{1,\,1,\,2}$ and $N = 1$ (only one orbital occupied). So, the list of states of system B that satisfy these requirements are:

List of states of system **B** satisfying:

$$U_B = U_{1,1,2} \text{ and } N_B = 1$$

$$\begin{pmatrix} u & o & u & u & u & u & u & u \\ 1,1,1 & 1,1,2 & 1,2,1 & 1,2,2 & 2,1,2 & 2,2,1 & 1,1,3 & \dots \end{pmatrix}$$

$$\begin{pmatrix} u & u & o & u & u & u & u & u \\ 1,1,1 & 1,1,2 & 1,2,1 & 1,2,2 & 2,1,2 & 2,2,1 & 1,1,3 & \dots \end{pmatrix}$$

Notice that the list excludes the orbital of interest, since, according to our conceptual division, system B contains all orbitals but the one of interest. If you compare this list with the one for the total system above, you will notice that it indeed coincides with the first two states, where the orbital of interest is occupied.

It is useful to express this result in terms of the multiplicity of system B, which I call g_B. In our example, we have $g_B(U - U_{2,\,1,\,1}, N - 1) = 2$ (recall that, in our example, $U = 2\,U_{1,\,1,\,2}$ and $N = 2$). Indeed, according to the list above, there are only two states of system B satisfying the complementary conditions $U_B = U_{1,\,1,\,2}$ and $N_B = 1$.

If the orbital of interest is unoccupied, on the other hand, then the complementary properties are $U_B = 2\,U_{1,\,1,\,2}$ and $N_B = 2$. As expected, we have four states satisfying these conditions. They are:

List of states of system **B** satisfying:

$$U_B = 2\,U_{1,1,2} \text{ and } N_B = 2$$

$$\begin{pmatrix} u & o & o & u & u & u & u & u \\ 1,1,1 & 1,1,2 & 1,2,1 & 1,2,2 & 2,1,2 & 2,2,1 & 1,1,3 & \cdots \end{pmatrix}$$

$$\begin{pmatrix} o & u & u & o & u & u & u & u \\ 1,1,1 & 1,1,2 & 1,2,1 & 1,2,2 & 2,1,2 & 2,2,1 & 1,1,3 & \cdots \end{pmatrix}$$

$$\begin{pmatrix} o & u & u & u & o & u & u & u \\ 1,1,1 & 1,1,2 & 1,2,1 & 1,2,2 & 2,1,2 & 2,2,1 & 1,1,3 & \cdots \end{pmatrix}$$

$$\begin{pmatrix} o & u & u & u & u & o & u & u \\ 1,1,1 & 1,1,2 & 1,2,1 & 1,2,2 & 2,1,2 & 2,2,1 & 1,1,3 & \cdots \end{pmatrix}$$

These are, indeed, the four states of system B that combine with system A when the orbital of interest is unoccupied. Thus, the multiplicity of system B when the orbital of interest is unoccupied is $g_B(U, N) = 4$, where, again, we recall that $U = 2U_{1,\,1,\,2}$ and $N = 2$ in our example.

This discussion is summarized in the list below, where I denote the number of states for which the orbital of interest is occupied and unoccupied, respectively, by $N_{occupied}$ and $N_{unoccupied}$.

$$g(2U_{1,1,2}, 2) = 6$$

Possible states:

$$\left.\begin{matrix} \begin{pmatrix} o & u & o & u & u & u & u & u & u \\ 2,1,1 & 1,1,1 & 1,1,2 & 1,2,1 & 1,2,2 & 2,1,2 & 2,2,1 & 1,1,3 & \cdots \end{pmatrix} \\ \begin{pmatrix} o & u & u & o & u & u & u & u & u \\ 2,1,1 & 1,1,1 & 1,1,2 & 1,2,1 & 1,2,2 & 2,1,2 & 2,2,1 & 1,1,3 & \cdots \end{pmatrix} \end{matrix}\right\} N_{occupied} = g_B(2U_{1,1,2} - U_{2,1,1}, N-1) = 2$$

$$\left.\begin{matrix} \begin{pmatrix} u & u & o & o & u & u & u & u & u \\ 2,1,1 & 1,1,1 & 1,1,2 & 1,2,1 & 1,2,2 & 2,1,2 & 2,2,1 & 1,1,3 & \cdots \end{pmatrix} \\ \begin{pmatrix} u & o & u & u & o & u & u & u & u \\ 2,1,1 & 1,1,1 & 1,1,2 & 1,2,1 & 1,2,2 & 2,1,2 & 2,2,1 & 1,1,3 & \cdots \end{pmatrix} \\ \begin{pmatrix} u & o & u & u & u & o & u & u & u \\ 2,1,1 & 1,1,1 & 1,1,2 & 1,2,1 & 1,2,2 & 2,1,2 & 2,2,1 & 1,1,3 & \cdots \end{pmatrix} \\ \begin{pmatrix} u & o & u & u & u & u & o & u & u \\ 2,1,1 & 1,1,1 & 1,1,2 & 1,2,1 & 1,2,2 & 2,1,2 & 2,2,1 & 1,1,3 & \cdots \end{pmatrix} \end{matrix}\right\} N_{unoccupied} = g_B(2U_{1,1,2}, N) = 4$$

Now we are ready to express the probability of finding the orbital of interest occupied in terms of the multiplicity of system B. Since we have been using the orbital 2, 1, 1 in the example, I will denote this probability by $P\left(\begin{smallmatrix} o \\ 2,1,1 \end{smallmatrix}\right)$. We know that this probability is just the ratio of the number of states where the orbital is occupied to the total number of states. Thus:

$$P\left(\begin{matrix} o \\ 2,1,1 \end{matrix}\right) = \frac{N_{occupied}}{N_{occupied} + N_{unoccupied}} = \frac{g_B\left(2U_{1,1,2} - U_{2,1,1}, N-1\right)}{g\left(2U_{1,1,2}, 2\right)} = \frac{2}{6}$$

and, also:

$$P\left(\begin{matrix} u \\ 2,1,1 \end{matrix}\right) = \frac{N_{unoccupied}}{N_{occupied} + N_{unoccupied}} = \frac{g_B\left(2U_{1,1,2}, N\right)}{g\left(2U_{1,1,2}, 2\right)} = \frac{4}{6}$$

Where $P\left(\begin{smallmatrix} u \\ 2,1,1 \end{smallmatrix}\right)$ is the probability of finding the orbital unoccupied. Obviously, it is also true that:

$$P\left(\begin{matrix} o \\ 2,1,1 \end{matrix}\right) + P\left(\begin{matrix} u \\ 2,1,1 \end{matrix}\right) = 1$$

That is the logic behind the calculation. Once it is grasped for a small system, it can easily be generalized for a large system. Indeed, if we had a macroscopic system, then $U = U_0$, $N = N_0$, where U_0 and N_0 are macroscopic quantities (N_0, for example, would be of the order of *moles*). So, one could envisage a super list, containing all the possible states of the macroscopic system, and that this list has again being conceptually divided into the orbital of interest (system A) and the others (system B). But because now we have lots and lots of electrons in system B, we should call it R, for reservoir, to remind ourselves that we are dealing with macroscopic quantities. Picking again the orbital 2, 1, 1 as the one of interest, the superlist would thus look like this:

$$g(U_0, \mathrm{N}_0) = N_{occupied} + N_{unoccupied}$$

Superlist of possible states of a macroscopic system:

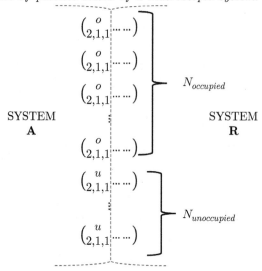

Of course, the logic to find the probability is still the same:

$$P\begin{pmatrix} o \\ 2,1,1 \end{pmatrix} = \frac{N_{occupied}}{N_{occupied} + N_{unoccupied}} = \frac{g_R(U_0 - U_{2,1,1}, N_0 - 1)}{g(U_0, \mathrm{N}_0)}$$

and:

$$P\begin{pmatrix} u \\ 2,1,1 \end{pmatrix} = \frac{N_{unoccupied}}{N_{occupied} + N_{unoccupied}} = \frac{g_R(U_0, N_0)}{g(U_0, \mathrm{N}_0)}$$

Now, to express the probability in a more convenient way, we are going to take advantage of the fact that we are considering a macroscopic system. We begin by dividing the two probabilities to get rid of the multiplicity of the total system:

$$\frac{P\begin{pmatrix} o \\ 2,1,1 \end{pmatrix}}{P\begin{pmatrix} u \\ 2,1,1 \end{pmatrix}} = \frac{g_R(U_0 - U_{2,1,1}, N_0 - 1)}{g_R(U_0, N_0)}$$

Now we express the multiplicity in terms of the entropy:

$$\frac{P\left(\begin{matrix} o \\ 2,1,1 \end{matrix}\right)}{P\left(\begin{matrix} u \\ 2,1,1 \end{matrix}\right)} = exp\left[\frac{S_R(U_0 - U_{2,1,1}, N_0 - 1) - S_R(U_0, N_0)}{k_B}\right]$$

System **R** is a reservoir, so we can again expand the entropy in a Taylor series and retain only the first order terms:

$$S_R(U_0 - U_{2,1,1}, N_0 - 1) \cong S_R(U_0, N_0) - U_{2,1,1}\frac{\partial S_R}{\partial U} - 1\frac{\partial S_R}{\partial N}$$

Using the definitions of temperature and potential:

$$S_R(U_0 - U_{2,1,1}, N_0 - 1) \cong S_R(U_0, N_0) - \frac{U_{2,1,1}}{T} + \frac{\mu}{T}$$

Thus:

$$\frac{P\left(\begin{matrix} o \\ 2,1,1 \end{matrix}\right)}{P\left(\begin{matrix} u \\ 2,1,1 \end{matrix}\right)} = exp\left[\frac{S_R(U_0, N_0) - \frac{U_{2,1,1}}{T} + \frac{\mu}{T} - S_R(U_0, N_0)}{k_B}\right]$$

that is:

$$\frac{P\left(\begin{matrix} o \\ 2,1,1 \end{matrix}\right)}{P\left(\begin{matrix} u \\ 2,1,1 \end{matrix}\right)} = exp\left[\frac{-U_{2,1,1} + \mu}{k_B T}\right]$$

This ratio gives an expression for $P\left(\begin{matrix} u \\ 2,1,1 \end{matrix}\right)$ in terms of $P\left(\begin{matrix} o \\ 2,1,1 \end{matrix}\right)$:

$$P\left(\begin{matrix} u \\ 2,1,1 \end{matrix}\right) = P\left(\begin{matrix} o \\ 2,1,1 \end{matrix}\right)exp\left[\frac{U_{2,1,1} - \mu}{k_B T}\right]$$

But, since $P\left(\begin{matrix} o \\ 2,1,1 \end{matrix}\right) + P\left(\begin{matrix} u \\ 2,1,1 \end{matrix}\right) = 1$, we have:

$$P\left(\begin{matrix} o \\ 2,1,1 \end{matrix}\right) + P\left(\begin{matrix} o \\ 2,1,1 \end{matrix}\right)exp\left[\frac{U_{2,1,1} - \mu}{k_B T}\right] = 1$$

and, therefore:

$$P\left(\begin{matrix} o \\ 2,1,1 \end{matrix}\right) = \frac{1}{1 + exp\left[\frac{U_{2,1,1}-\mu}{k_B T}\right]}$$

This is the probability of finding the orbital of interest expressed in terms of the energy of the orbital and parameters of the reservoir (the temperature and potential). Recall that the reservoir is the system "macroscopic number of electrons in a box" itself, excluding the orbital of interest, but since the system is macroscopic, exclusion of only one orbital does not alter its temperature and chemical potential (more details about the role of the reservoir are given in section 1.9). Obviously, we could have picked any orbital to do the calculation, so this expression is valid for any orbital. Thus, the probability of finding any given orbital c occupied is:

$$P\left(\begin{matrix} o \\ c \end{matrix}\right) = \frac{1}{1 + exp\left[\frac{U_c-\mu}{k_B T}\right]} \tag{1.65}$$

where U_c is the energy of the orbital c.

This important result is known as the Fermi–Dirac distribution. We have seen that the Fermi–Dirac distribution gives the probability of finding an orbital occupied in a macroscopic system, that is, in a system with a macroscopic number of electrons. Since the Fermi–Dirac distribution depends on the energy of the orbital, it is often expressed as a function of energy, as shown in Equation (1.66) below:

$$f(\varepsilon) = \frac{1}{1 + \exp\left[\frac{(\varepsilon-\mu)}{k_B T}\right]} \tag{1.66}$$

This equation is the same as Equation (1.65), just the notation was changed. I very much prefer the notation of Equation (1.65) though, because it reminds us that the probability refers to a specific orbital, not to a specific energy (if, for example, two orbitals have the same energy, then the probability of finding an electron with this energy is two times greater than the probability for each orbital, but the notation of Equation (1.66) obscures this issue: it looks like it is giving the probability for a specific energy, instead of a specific orbital). The notation of Equation (1.66), however, is far more common in the literature.

Notice that the logic we followed essentially treats the orbital of interest as a system in thermal and diffusive contact with the reservoir (it is in diffusive contact because the electron can leave the orbital of interest and go to another orbital of the reservoir). As such, we could have applied Equation (1.50) directly to the problem. Let us do that now. In the case of the system "orbital of interest", there are only two possible states: the orbital is either occupied or unoccupied. If it is occupied then $n = 1$ (recall that n is the number of particles in the system) and $\varepsilon = \varepsilon_c$, where ε_c is the energy of the orbital. If it is unoccupied, then $n = 0$ and $\varepsilon = 0$. Thus, the Gibbs sum, which for the system "orbital of interest" involves only two states, becomes:

$$\varsigma(T, \mu) = \exp\left(-\frac{0 - 0\mu}{k_B T}\right) + \exp\left(-\frac{\varepsilon_C - 1\mu}{k_B T}\right) = 1 + \exp\left(-\frac{\varepsilon_C - \mu}{k_B T}\right)$$

The Gibbs sum can now be used to calculate the probability of the system "orbital of interest" to be in the state "occupied". According to Equation (1.50):

$$P\binom{o}{c} = \frac{\exp\left[\frac{-(\varepsilon_c - n_c\mu)}{k_B T}\right]}{\varsigma(T, \mu)} = \frac{\exp\left[\frac{-(\varepsilon_c - \mu)}{k_B T}\right]}{1 + \exp\left[\frac{-(\varepsilon_c - \mu)}{k_B T}\right]}$$

$$\therefore P\binom{o}{c} = \frac{1}{1 + \exp\left[\frac{(\varepsilon_c - \mu)}{k_B T}\right]} \tag{1.67}$$

Equation (1.67) is, of course, identical to Equation (1.65).

Now let us try to better understand what the Fermi–Dirac distribution is teaching us. First, imagine that we freeze the system towards the absolute zero ($T \to 0\ K$). In this case, if $\varepsilon > \mu$, then the argument of the exponential in the denominator of Equation (1.66) is of the form "positive number divided by $T \to 0$", which means that, as $T \to 0$, then $\exp\left[\frac{(\varepsilon - \mu)}{k_B T}\right] \to$ $\exp[+\infty] \to \infty$, thus resulting in $f(\varepsilon) = 0$ for $\varepsilon > \mu$. On the other hand, if $\varepsilon < \mu$, then the argument of the exponential is of the form "negative number divided by $T \to 0$", which means that, as $T \to 0$, then $\exp\left[\frac{(\varepsilon - \mu)}{k_B T}\right] \to$ $\exp[-\infty] \to 0$, thus resulting in $f(\varepsilon) = 1$ for $\varepsilon < \mu$. Conclusion: at the absolute zero, all orbitals with energies lower than μ are occupied, while all orbitals with energies higher than μ are unoccupied. This situation is shown in the blue line of Figure 1.7, where the Fermi–Dirac distribution is plotted as function of the energy assuming $\mu = 1\ e.\ V$. Notice that there is a sharp transition from probability one to probability zero at the point $\varepsilon = \mu$.

As the temperature increases, this transition is softened, as shown in the Fermi–Dirac distribution for room temperature (green line of Figure 1.7).

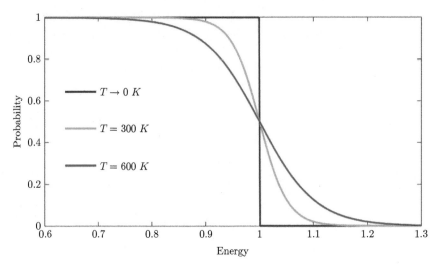

Figure 1.7 Plot of Fermi–Dirac distribution against energy assuming $\mu = 1$ for three different temperatures: $T \to 0$ (blue line), $T = 300$ K (green line), $T = 600$ K (red line).

Now, there is a probability that some electrons occupying orbitals with energy lower than μ will be excited to orbitals with energies higher than μ. Such an excitation is due to the thermal energy pumped into the system by increasing its temperature. As the temperature increases even further, the probability of occupation of orbitals with lower energies decreases even further (compare the red and green lines in the region $\varepsilon < \mu$), while the probability of occupation of orbitals with higher energy increases even further (compare the red and green lines in the region $\varepsilon > \mu$). Thus, the Fermi–Dirac distribution captures the effect of thermal energy, which is to excite electrons from orbitals with lower energies to orbitals with higher energy.

As a matter of fact, at the absolute zero, the potential μ receives a special name: it is called the Fermi level. Denoting it by ε_F, we thus have, by definition:

$$\mu(T = 0) = \varepsilon_F \tag{1.68}$$

Unfortunately, there is a huge inconsistency in the literature when using the term "Fermi level". Though its "official" definition is given by Equation (1.68), more often than not one comes across books and scientific articles using this term as a synonym for the total potential μ (in the context of semiconductors, the potential μ is almost always the electrochemical potential). In other words, the literature in semiconductors generalize the definition of Equation (1.68) for any temperature. This has become so common that, in the following chapters, I will bow down to the new convention and also use it as a synonym for the electrochemical potential μ. But beware of this difference when studying the literature!

Finally, notice that we can also interpret the electrochemical potential μ (or the Fermi level ε_F, using the new convention), as being the energy for which the probability of occupation is equal to 0.5.

1.8 Analogy between the systems "box" and "coins"

The physical system "electrons in a box", which is the system of interest in the field of semiconductors, is analogous to the system "coins", which we used a couple of times to illustrate the logic of statistical physics. In this section, I compare both systems in more detail, to facilitate the transition between the pedagogical system "coins" and the physical system "electrons in a box".

First, notice that, strictly speaking, it is not the electrons that play the role of the coins, but the orbitals. Likewise, it is the states "occupied" and "unoccupied" that play the role of the states *heads* and *tails*.

Both coins and orbitals are distinguishable. To illustrate this fact, suppose that we have two coins, say $M1$ and $M2$; analogously, we have two orbitals, say the orbitals 1, 1, 1 and 1, 1, 2. The coins are distinguishable, which means that the state $\begin{pmatrix} H & T \\ M1 & M2 \end{pmatrix}$ is different from the state $\begin{pmatrix} T & H \\ M1 & M2 \end{pmatrix}$:

$$\begin{pmatrix} H & T \\ M1 & M2 \end{pmatrix} \neq \begin{pmatrix} T & H \\ M1 & M2 \end{pmatrix}$$

Analogously, the orbitals are distinguishable, which means that the state $\begin{pmatrix} u & o \\ 1,1,1 & 1,1,2 \end{pmatrix}$ is different from the state $\begin{pmatrix} o & u \\ 1,1,1 & 1,1,2 \end{pmatrix}$:

$$\begin{pmatrix} u & o \\ 1,1,1 & 1,1,2 \end{pmatrix} \neq \begin{pmatrix} o & u \\ 1,1,1 & 1,1,2 \end{pmatrix}$$

The electrons, on the other hand, are indistinguishable. This means that the state $\begin{pmatrix} u & o \\ 1,1,1 & 1,1,2 \end{pmatrix}$ is identical to the state $\begin{pmatrix} o & u \\ 1,1,2 & 1,1,1 \end{pmatrix}$. Indeed, these are two equivalent ways of expressing the same state: both represent the state "orbital 1, 1, 1 unoccupied and orbital 1, 1, 2 occupied". Thus:

$$\begin{pmatrix} u & o \\ 1,1,1 & 1,1,2 \end{pmatrix} = \begin{pmatrix} o & u \\ 1,1,2 & 1,1,1 \end{pmatrix}$$

The analogy of the indistinguishability of electrons in the system "coins" is the fact that the order of the coins does not matter: it does not matter if I put the coin $M1$ first or second in the list. Indeed, the state $\begin{pmatrix} H & T \\ M1 & M2 \end{pmatrix}$ is identical to the state $\begin{pmatrix} T & H \\ M2 & M1 \end{pmatrix}$. Thus

$$\begin{pmatrix} H & T \\ M1 & M2 \end{pmatrix} = \begin{pmatrix} T & H \\ M2 & M1 \end{pmatrix}$$

To avoid confusion, recall that:

$$\begin{pmatrix} H & T \\ M1 & M2 \end{pmatrix} \neq \begin{pmatrix} T & H \\ M1 & M2 \end{pmatrix}$$

Hitherto the analogy has been perfect, but it breaks down a bit in the energy. Indeed, in our example of the coins, we could have only two possible energies ascribed to each coin, but we know that the orbitals can have a wide range of energies associated with them (as a matter of fact, in principle we have an infinite number of energies, since any combination of natural numbers is possible to designate an orbital). This is, however, the only difference between the two systems.

1.9   Concentration of electrons and Fermi level

Typically, books on semiconductors assume that the reader is acquainted with the Fermi–Dirac distribution. Thus, this is usually the starting point to derive the electrical properties of semiconductor devices. This procedure will be shown in the following chapters. For now, we need to get a good grasp on the physical meaning of the Fermi–Dirac distribution. We saw that it is the probability of finding a given orbital occupied. Now we show how the parameters used in the previous section translate into parameters of semiconductors.

First, notice that the "box" we have been talking about will be a piece of semiconductor, say silicon. The energy U_0, on the other hand, depends on the temperature of the environment where the semiconductor is: the lab, for example. Indeed, the energy U_0 is the one that satisfy the condition:

$$\left. \frac{\partial S_{semiconductor}}{\partial U} \right|_{U=U_0} = \frac{1}{T} \tag{1.69}$$

where T is the temperature of the environment.

As in the example of the "box", N_0 is still the total number of electrons in the semiconductor (as we will see in the following chapters, we are particularly interested in the free electrons, that is, electrons that can contribute to the current). Thus, the Fermi–Dirac distribution can be used to relate the total number of electrons N_0 to the potential μ. Since each orbital c has a probability $P\left(\begin{smallmatrix} o \\ c \end{smallmatrix}\right)$ of being occupied, then:

$$N_0 \cong \sum_c P\left(\begin{smallmatrix} o \\ c \end{smallmatrix}\right) \tag{1.70}$$

Notice that the sum in Equation (1.70) is over all possible orbitals. We will return to this equation in Chapter 2, and explain in more detail in what sense it is true. Here, I just want to emphasize that the Fermi–Dirac distribution, through Equation (1.70), relates the electrochemical potential μ (or, equivalently, the Fermi level ε_F) to the number of electrons N_0. That makes sense, because the potential μ involves the chemical potential, which, as we saw earlier, is connected with the concentration of particles. Equation (1.70) can be used to connect them explicitly, as shall be done in Chapter 2.

1.10 Transport

This section deals with phenomena of particle transport, that is, with how particles move. We are particularly interested in studying transport of charges, which are usually expressed in terms of currents, but the derivations in this section are general and apply to any kind of particle, be it charged or not. The two most important types of currents are the drift current and the diffusion current. The drift current is the simplest of the two: it is the current that flows through a resistor when an electric field (or, equivalently, an electrostatic potential difference) is applied to it. We begin by considering the drift current, and then move on to the diffusion current, which is less simple, and as such will take the largest part of this section.

Consider a tube filled with a gas of particles, as shown in Figure 1.8 (it does not matter whether they are charged or not). We want to know what

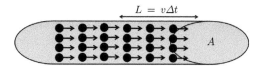

$L = v\Delta t$

A

Figure 1.8 Drift current in a tube with cross-sectional area A.

the current of particles through the cross-sectional area is, that is, how many particles cross the area per unit of time. Specifically, we want an expression for the current in terms of the velocity and volume concentration of particles. Suppose that all particles have the same speed v, and that the direction of velocity points to the right. As can be seen in Figure 1.8, all particles inside the volume delimited by the length $L = v\Delta t$ will have crossed the sectional area A in an interval of time Δt. Denoting the volume concentration of particles by ρ, then, the total number of particles $N_{\Delta t}$ that will have crossed the area A in an interval of time Δt is:

$$N_{\Delta t} = \rho \cdot L \cdot A = \rho \cdot v \cdot \Delta t \cdot A \tag{1.71}$$

The current I is the number of particles that crosses the area per unit of time. Thus:

$$I = \frac{N_{\Delta t}}{\Delta t} = \rho \cdot v \cdot A \tag{1.72}$$

It is more convenient, however, to work with the current density J, that is, the current per unit of cross-sectional area ($J = I/A$). Thus, we conclude that the drift current density J_{drift} is:

$$J_{drift} = \rho \cdot v \tag{1.73}$$

Equation (1.73) is the simplest form of the drift current density. Notice that the physical origin of this current is the preferential velocity of the particles (they all point to the right). Thus, the drift current is associated with a coordinated movement of particles. Notice, furthermore, that Equation (1.73) does not really describe a physical law: it is only a logical relationship (given the definition of the three terms, the relation follows logically). The physics comes in when we relate the velocity to an external force. In the case of the drift current, this relation is described by Ohm's law, which essentially affirms that the velocity is proportional to the electric field. We will use Ohm's law in the following chapters, where we consider transport of charges in semiconductor devices.

Now we generalize the expression for the drift current by considering the possibility that the particles may have different velocities. For the sake of simplicity, suppose that the direction of all velocities are parallel to the horizontal axis, but that their magnitude and sign may change. To work out the current density, suppose that there are N groups of particles, and that all particles inside each group have the same velocity (but particles in different groups can have different velocities). If all particles inside each group have the same velocity, then we can use Equation (1.73) to compute the contribution of each group to the current and add them up to find the total current.

We denote the concentration of particles of a given group by ρ_n and their velocity by v_n (which can be either positive or negative). Thus, each group contribute with $\rho_n \cdot v_n$ to the total current, which is found by adding up all contributions:

$$J_{drift} = \sum_{n=1}^{N} \rho_n \cdot v_n \tag{1.74}$$

We can recast Equation (1.74) in a more convenient form by noticing that, if the group n has N_n particles, then the mean velocity $\langle v \rangle$ of all particles is:

$$\langle v \rangle = \frac{\sum_{n=1}^{N} N_n \cdot v_n}{\sum_{n=1}^{N} N_n} \tag{1.75}$$

Denoting the volume by V, then:

$$\langle v \rangle = \frac{\sum_{n=1}^{N} V \cdot \rho_n \cdot v_n}{\sum_{n=1}^{N} V \cdot \rho_n} = \frac{\sum_{n=1}^{N} \rho_n \cdot v_n}{\sum_{n=1}^{N} \rho_n}$$

$$\therefore \langle v \rangle = \frac{\sum_{n=1}^{N} \rho_n \cdot v_n}{\rho} \tag{1.76}$$

where ρ is the total concentration of particles.

With the help of Equation (1.76), Equation (1.74) can be recast as:

$$J_{drift} = \rho \langle v \rangle \tag{1.77}$$

It is thus apparent that, if the mean velocity of particles is zero, then the drift current is also zero. Thus, we can give a more precise meaning to the notion of a "preferential velocity": there is a preferential velocity when the mean velocity is different from zero. This preferential velocity is induced by a force, as described by Ohm's law.

Now we move on to a description of the diffusion current. Contrary to the drift current, the diffusion current is not due to a preferential velocity. Indeed, to describe the diffusion current, we assume that $\langle v \rangle = 0$. Since it is a force that causes $\langle v \rangle$ to be different from zero, then there is no force associated with a diffusion current: it exists in the absence of forces!

A diffusion current arises when the concentration of particles is not uniform. Thus, the diffusion current is associated with a variation in the chemical potential. To find an expression for the diffusion current, consider two boxes of length l_z, as shown in Figure 1.9. We will shortly discuss the physical meaning of l_z. For now, though, just take it as the length of the box. We

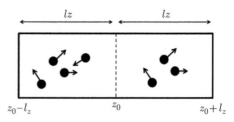

Figure 1.9 Schematics to calculate the diffusion current.

assume that the mean velocity is zero, but that the two boxes have a different concentration of particles (if there is both a difference in concentration and a mean velocity different from zero, then there is both a diffusion and a drift current, and the total current is the sum of both; thus, when treating the diffusion current, we can always assume that the mean velocity is zero and include its effect in the drift current). We also assume that all particles have the same magnitude of velocity, that is, the same speed, but that the direction of the velocity is random, thus resulting in $\langle v \rangle = 0$.

Our job is to compute the diffusion current in terms of the variation in the concentration. The diffusion current is still a current, so it is still the number of particles that crosses a cross-sectional area per unit of time. We will compute the diffusion current at the interface between the two boxes (point z_0 in Figure 1.9). We denote the current from left to right by J_{l-r} and the current from right to left by J_{r-l}. All we need to do is to find both contributions and subtract one from the other to find the diffusion current.

To compute J_{l-r} and J_{r-l}, we can use a similar reasoning to the one leading to Equation (1.77): we can divide the particles into groups, each group containing particles with velocity in the same direction, compute the contribution of each group using Equation (1.73), and then add them all up (notice that we can use Equation (1.73) for each group separately because all particles in each group have the same velocity). However, we need to adapt Equation (1.73) to include velocities that are not parallel to the horizontal axis (to the z-axis of Figure 1.9), but that is easy: all we need to do is to take the z-component of the velocity.

There is, however, a more subtle effect that we also need to consider: the reasoning leading to Equation (1.73) clearly assumes that the velocities do not change as the particles move towards the cross-sectional area. In the physical situation described in Figure 1.9, however, we need to consider that the particles may collide, among themselves, but mainly with the atomic structure of the material through which the current flows. In fact, collision with the atomic structure also happens in the drift current, but this effect is embedded in the linear relationship between the velocity and the force (that is, in Ohm's law). That is why we did not need to worry about collision in the reasoning that led to Equation (1.73). Does that mean that Equation (1.73) cannot be used to find the contribution of each group of particles? Of course not, we can still use Equation (1.73), but we need to be careful to limit its application to a situation where there are no collisions. So, all we need to do is to make sure that our boxes are sufficiently small to guarantee that there will be no collision when the particle is travelling through the box towards the intersection. So, suppose that we have a given material for which we want to calculate the diffusion current. We then zoom in a certain region of the material, which becomes the boxes of Figure 1.9. Now we ask the question: what is the typical distance that a particle inside this material travels before it collides? The parameter that gives this distance is the mean free path. So, it is the mean free path that tells us what the size of the box should be, because if the boxes lengths coincide with the mean free path, then there will be no collision between particles, thus allowing us to use Equation (1.73) to compute the contribution of each group of particles. Notice that, on the one hand, we cannot take the boxes to be shorter than the mean free path, because then we would not be including all particles that do contribute to the current; on the other hand, if we took it to be larger, we would be including particles that do not contribute to the current. Conclusion: their lengths must coincide with the mean free path.

Ah, but there is still one more issue we need to consider: if a group of particles have a velocity that is not parallel to the horizontal axis (the z-axis in Figure 1.9), then the horizontal distance it will travel is not the mean free path, but the projection of the mean free path onto the z-axis. Suppose that we are analysing a group of particles whose velocity makes an angle θ with the z-axis. Then, the z-component of the velocity is given by:

$$v_z = \left|\vec{v}\right| \cos\theta \tag{1.78}$$

Likewise, denoting the mean free path by l, if the direction of the velocity of the group of particles makes an angle θ with the z-axis, then the horizontal distance travelled by the particles in this group is:

$$l_z = l \cos\theta \tag{1.79}$$

Thus, the length of the box for this group of particles must be $l_z = l \cos\theta$.

Now we are ready to use Equation (1.73) to compute the contribution of each group of particles to the total current. We begin by computing J_{l-r}. Since the particles that contribute to J_{l-r} come from the left box, we need to use the concentration of particles in the left box. We need to treat the concentration as a function of z (since it varies across the material), and as a function of θ, since we split the total number of particles into groups according to the direction of their velocity. Thus, the concentration of particles is a function of two variables, and we denote it by $\rho(z, \theta)$. In particular, the concentration in the left box for the group travelling with an angle θ is $\rho(z_0 - l_z, \theta)$, so the contribution of this group of particles to J_{l-r} is:

$$\Delta J_{l-r}(\theta) = \rho(z_0 - l_z, \theta) \cdot v_z \qquad (1.80)$$

The procedure to find the contribution to J_{r-l} is the same, but now we need to use the concentration of particles in the right box. Thus:

$$\Delta J_{r-l}(\theta) = \rho(z_0 + l_z, \theta) \cdot v_z \qquad (1.81)$$

The contribution to the total diffusion current of the group of particles with angle θ is found by subtracting $\Delta J_{r-l}(\theta)$ from $\Delta J_{l-r}(\theta)$ (the order of the subtraction depends on where the z-axis points to, and, in Figure 1.9, z-positive points to the right):

$$\Delta J(\theta) = \Delta J_{l-r}(\theta) - \Delta J_{r-l}(\theta)$$

$$\therefore \Delta J(\theta) = [\rho(z_0 - l_z, \theta) - \rho(z_0 + l_z, \theta)] \cdot v_z \qquad (1.82)$$

Typically, the concentration of particles varies slowly with respect to the mean free path length, which allows treating l_z as an infinitesimal quantity. Thus, in this condition, it follows that:

$$\frac{\rho(z_0 + l_z, \theta) - \rho(z_0 - l_z, \theta)}{2 \cdot l_z} \approx \frac{\partial \rho}{\partial z}$$

$$\therefore \rho(z_0 - l_z, \theta) - \rho(z_0 + l_z, \theta) \approx -\frac{\partial \rho}{\partial z} 2 \cdot l_z \qquad (1.83)$$

Therefore:

$$\Delta J(\theta) = -\frac{\partial \rho(z, \theta)}{\partial z} 2 \cdot v_z \cdot l_z \qquad (1.84)$$

Notice that Equation (1.83) implies that the derivative is evaluated at the interface between the boxes, that is, at $z = z_0$, which is precisely where we are calculating the current at.

Substituting Equation (1.78) and Equation (1.79) into Equation (1.84):

$$\Delta J(\theta) = -\frac{\partial \rho(z, \theta)}{\partial z} 2 \cdot |\vec{v}| \cdot l \cdot \cos^2 \theta \qquad (1.85)$$

Since the particles have random orientation, there is no preferential angle, that is, the particles are uniformly distributed among all angles. Thus, the concentration of particles that point in the direction of a solid angle $d\Omega = \sin(\theta) d\theta d\varphi$ is:

$$\rho(z, \theta) = \rho(z) \frac{d\Omega}{4\pi} \qquad (1.86)$$

Where $\rho(z)$ is the total concentration of particles and 4π is the integral of the solid angle over a closed sphere (or, if you prefer, the area of a sphere with radius 1). Notice that Equation (1.86) is just asserting that the fraction of the particles that points towards the solid angle $d\Omega$ is equal to the fraction $d\Omega/4\pi$, as expected for a uniform distribution. With the help of Equation (1.86), Equation (1.85) can be expressed as:

$$\Delta J(\theta) = -\frac{\partial \rho(z)}{\partial z} 2 \cdot |\vec{v}| \cdot l \cdot \cos^2\theta \cdot \frac{d\Omega}{4\pi} \tag{1.87}$$

To find the total current, we need to add up the contributions of all group of particles, that is, we need to integrate Equation (1.87) over a half-sphere (we integrate over a half-sphere, and not a full sphere, because only one half contributes to the current: neither the particles in the left box whose velocities point to negative z nor the particles in the right box whose velocities point to positive z contribute to the current). Thus:

$$J_{dif} = \int \Delta J(\theta) = \frac{-\frac{\partial \rho(z)}{\partial z} 2 \cdot |\vec{v}| \cdot l}{4\pi} \int \cos^2\theta \, d\Omega$$

$$\therefore J_{dif} = \frac{-\frac{\partial \rho(z)}{\partial z} 2 \cdot |\vec{v}| \cdot l}{4\pi} \int_0^{2\pi} \int_0^{\pi/2} \cos^2\theta \sin\theta \, d\theta \, d\varphi \tag{1.88}$$

Notice, once again, that we are integrating over a half-sphere (that is, the integral over the variable θ goes from 0 to $\pi/2$). The integral offers no difficulties, and the result is:

$$J_{dif} = -\frac{1}{3} \cdot |\vec{v}| \cdot l \cdot \frac{\partial \rho}{\partial z} \tag{1.89}$$

Thus, we conclude that the diffusion current density is proportional to the concentration of particles gradient in the direction we are assessing the current (the z direction here). This is not a surprising result, since from the beginning we knew that it is the change in the concentration that causes the diffusion current. The constant of proportion is called the diffusion coefficient D. The diffusion current density is usually expressed in terms of D, as:

$$J_{dif} = -D \cdot \frac{\partial \rho}{\partial z} \tag{1.90}$$

Thus, according to our model:

$$D = \frac{1}{3} \cdot |\vec{v}| \cdot l \tag{1.91}$$

Since we considered the current in the z direction, then Equation (1.90) is the z component of the current. But the procedure to find the other components is the same, just replacing the axis. Thus, the total current (which is a vector) is:

$$\overrightarrow{J_{dif}} = J_x \hat{x} + J_y \hat{y} + J_z \hat{z} \tag{1.92}$$

Where the components are given by:

$$J_x = -D \cdot \frac{\partial \rho}{\partial x}, J_y = -D \cdot \frac{\partial \rho}{\partial y}, J_z = -D \cdot \frac{\partial \rho}{\partial z} \tag{1.93}$$

Notice that we can conveniently express the diffusion current density in terms of the gradient of concentration of particles:

$$\overrightarrow{J_{dif}} = -D\nabla\rho \tag{1.94}$$

Equation (1.94) is a more general form for the diffusion current density, but for our purposes the one-dimensional form (Equation (1.90)) is sufficient. It is worth pointing out that other models may arrive at different expressions for the diffusion coefficient D (for example, anisotropic materials can have different diffusion coefficients in different directions). The most important bit, however, is to notice that the diffusion current is proportional to the spatial variation of the concentration of particles. What exactly the proportion is depends on the physical conditions and the specific materials. Such a relation is known as Fick's law.

Finally, notice that the negative sign between the diffusion current and the concentration gradient comes from the fact that, if we have a higher concentration in the left box, then the current is from left to right (positive z). But a higher concentration in the left box means that the derivative of the concentration is negative (it diminishes in the direction of positive z). Thus, the negative sign is telling us that the current is in the direction of lower concentration, which means that the current acts to homogenize the concentration (and, consequently, the chemical potential), as expected.

What would happen to the universe if a mischievous daemon flipped the sign in Equation (1.94)?

1.11   Relationship between current and concentration of particles (continuity equation)

Our last job in this long but important chapter is to find an expression that connects the currents to the concentration of particles, the so-called "continuity equation". The equations derived here will be used later in the derivation of the relationship between current and voltage in a **p-n** junction (the famous Shockley equation).

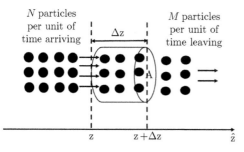

N particles
per unit of
time arriving

Δz

M particles
per unit of
time leaving

A

z   z+Δz $\hat{z}$

Figure 1.10 A box of length Δz and area A. There is a current of N particles per unit time arriving in the box, and M particles per unit of time leaving the box.

Suppose we have a box of length Δz and area A, as shown in Figure 1.10, through which there is a current flowing. Our goal is to relate the current flow with the particle concentration inside the box. The logic is very simple: suppose that there is a current of N particles per unit time arriving in the box, and a current of M particles per unit time leaving the box. If there are N particles per unit time arriving, then in an interval of time Δt there will be an increase of $N\Delta t$ particles inside the box. Likewise, there will be a decrease of $M\Delta t$ particles in the interval Δt. We denote the total number of particles inside the box by a function of time $P(t)$. Thus, it follows that:

$$(N - M)\Delta t = P(t + \Delta t) - P(t) \tag{1.95}$$

where $P(t + \Delta t) - P(t)$ is the increase in the number of particles in the interval Δt. This is all there is to the logic: we just expressed in mathematical terms the notion that "what enters, minus what leaves, is what is left". A very sophisticated business indeed!

All we need to do now is to massage Equation (1.95) into a more convenient form. First, instead of ascribing an incoming current N and an outgoing current M, it is much more convenient to express the current as a function of the position and identify N and M with the value of the current in the two respective positions. Thus, denoting the current by $I(z)$ and placing a coordinate system such that the beginning of the box is at z and the end at $z + \Delta z$, we readily identify that $N = I(z)$ and $M = I(z + \Delta z)$. Using this notation, Equation (1.95) reads:

$$[I(z) - I(z + \Delta z)]\Delta t = P(t + \Delta t) - P(t) \tag{1.96}$$

We can improve the elegance of our little equation by expressing the total number of particles in terms of the concentration. As it is obvious in Figure 1.10, the volume of the box is $V = A\Delta z$. Denoting again the concentration of particles by ρ, it follows that $P = \rho A\Delta z$. Thus, Equation (1.96) can be recast as:

$$[I(z) - I(z + \Delta z)]\Delta t = [\rho(t + \Delta t) - \rho(t)]A\Delta z$$

As mentioned in the previous section, it is often more convenient to work with the current density J instead of the total current I. From the definition of both terms, it follows immediately that $I(z) = J(z)A$, thus:

$$[J(z) - J(z + \Delta z)]A\Delta t = [\rho(t + \Delta t) - \rho(t)]A\Delta z$$

Rearranging:

$$\frac{J(z) - J(z + \Delta z)}{\Delta z} = \frac{\rho(t + \Delta t) - \rho(t)}{\Delta t} \tag{1.97}$$

If we take the length of the box and the interval of time to be infinitesimal, then Equation (1.97) becomes:

$$-\frac{\partial J}{\partial z} = \frac{\partial \rho}{\partial t} \tag{1.98}$$

Notice that Equation (1.98) involves partial derivatives, because J and ρ may be functions of both time and space. Notice, moreover, the negative sign on the left-hand side of the equation, which appears because the numerator on the left-hand side of Equation (1.97) is $J(z) - J(z + \Delta z) = -[J(z + \Delta z) - J(z)]$. The negative sign makes sense: if $J(z) > J(z + \Delta z)$, than there are more particles arriving than leaving, which means that the concentration of particles is increasing with time. But $J(z) > J(z + \Delta z)$ also implies a negative partial derivative ($J(z)$ decreases when moving from z to $z + \Delta z$). Furthermore, notice that we have not specified what kind of current J is: it can be drift, diffusion or a combination of both.

It is important to avoid mystifying Equation (1.98). Recall that it is just the mathematical expression of the notion that "what enters, minus what leaves, is what is left".

In the following chapters, we will need to describe the current of two types of particles: negative particles (good old electrons), and positive particles (a new kid on the block, called "holes", which will be explained in the next chapter). Thus, it is convenient to express Equation (1.98) in terms of these two types of particles. We denote the concentration of positively charged particles by the letter p, and the concentration of negatively charged particles by the letter n.

These two quantities have units of inverse of volume, because they are just the number of particles per unit of volume. An electric current, however, is usually expressed in terms of concentration of charges, that is, in units of coulombs per volume. Since each positive particle has a charge of $|q|$ coulombs, then the concentration of positive charges is $\rho = |q|\, p$. Thus, for positively charged particles, Equation (1.98) can be cast as:

$$-\frac{\partial J}{\partial z} = |q|\frac{\partial p}{\partial t} \tag{1.99}$$

Likewise, the concentration of negative charges is $\rho = - \mid q \mid n$. Thus, for negatively charged particles, Equation (1.98) can be cast as:

$$-\frac{\partial J}{\partial z} = - \mid q \mid \frac{\partial n}{\partial t} \qquad (1.100)$$

Therefore, when dealing with holes, we will use Equation (1.99), and when dealing with electrons we will use Equation (1.100).

In the field of semiconductor devices, we are often interested in particles that can contribute to electric currents. As we will see in the next section, both electrons and holes can suffer a process called "recombination", in which they cease to be a free particle, that is, a particle that can contribute to currents. Thus, the recombination process induces a loss in the concentration of free particles, that is, in n and p, which requires that we modify Equation (1.99) and Equation (1.100) to include this effect. Recall, once again, that these two equations are mathematical expressions of the notion that "what enters, minus what leaves, is what is left". All we need to do is to extend this notion to include losses, that is, extend it into "what enters, minus what leaves, minus what dies, is what is left".

To do this, let us go back to Equation (1.97), which I rewrite below in a more convenient form:

$$[J(z) - J(z + \Delta z)]A\Delta t = [\rho(t + \Delta t) - \rho(t)]A\Delta z$$

Recall that the left-hand side of this equation tells us what is entering minus what is leaving, while the right-hand side tells us what stays. Now we need to include what dies. For that, we need a parameter that quantifies the losses.

The losses are quantified in terms of a frequency γ. This parameter is the probability per unit time that the particles will recombine. Thus, if there are $P(t)$ particles in the volume, then $P(t)\gamma$ particles recombine per unit of time, which means that $P(t)\gamma\Delta t$ particles die in an interval of time Δt. This expression can be used to incorporate the contribution of recombination losses into the left-hand side of Equation (1.97), which now reads:

$$[J(z) - J(z + \Delta z)]A\Delta t - \rho(t)A\Delta z\gamma\Delta t = [\rho(t + \Delta t) - \rho(t)]A\Delta z$$

where $P(t) = \rho(t)A\Delta z$ was used.

Dividing both sides by $A\Delta z\Delta t$:

$$\frac{[J(z) - J(z + \Delta z)]}{\Delta z} - \rho(t)\gamma = \frac{[\rho(t + \Delta t) - \rho(t)]}{\Delta t}$$

For infinitesimal quantities, the equation above becomes:

$$-\frac{\partial J}{\partial z} - \rho\gamma = \frac{\partial\rho}{\partial t} \tag{1.101}$$

Equation (1.101) is the version of Equation (1.98) that includes recombination losses. The frequency γ is often expressed in terms of the lifetime τ, defined as:

$$\tau = \frac{1}{\gamma} \tag{1.102}$$

Thus, the lifetime τ is the typical time that it takes for a particle to recombine (or, if you prefer, to die – see Box 6).

Box 6   The physical meaning of the lifetime

In order to gain insights into the physical meaning of the lifetime, let us consider a situation where we have a concentration of particles ρ subject to a loss process characterized by the lifetime τ. Furthermore, suppose that the current of particles is spatially uniform, so there is no contribution to changes in the concentration due to currents. The current and concentrations are related by:

$$-\frac{\partial J}{\partial z} - \frac{\rho}{\tau} = \frac{\partial\rho}{\partial t}$$

If the current is spatially uniform, then $\frac{\partial J}{\partial z} = 0$, so:

$$\frac{\partial\rho}{\partial t} = -\frac{\rho}{\tau}$$

Which entails that:

$$\rho(t) = \rho(0)e^{-\frac{t}{\tau}}$$

Thus, the concentration decays exponentially in time. How fast it decays is defined by the lifetime. Notice that:

$$\rho(\tau) = \rho(0)e^{-\frac{\tau}{\tau}} = \frac{\rho(0)}{e}$$

This means that, after τ seconds, the concentration is already reduced by a factor of e^{-1}. Thus, the lifetime is a measure of how long it takes for the particles to disappear due to the loss mechanism.

Equation (1.101) can now be expressed in terms of the concentration of holes:

$$-\frac{\partial J}{\partial z} - |q|\frac{p}{\tau_p} = |q|\frac{\partial p}{\partial t} \tag{1.103}$$

or the concentration of free electrons:

$$-\frac{\partial J}{\partial z} + |q|\frac{n}{\tau_n} = -|q|\frac{\partial n}{\partial t} \tag{1.104}$$

Notice, once again, that these equations are valid for all types of currents (drift, diffusion or a combination of both).

Now we focus attention on a specific application of these equations, which is of particular importance to our purposes. Suppose that there is a certain current, it does not matter which, that flows through the region $z < 0$, as shown in Figure 1.11. To be specific, we assume that it is a current of holes, but the logic is the same for electrons. Now suppose that, for $z \geq 0$, the current is only by diffusion. Our goal is to find an expression for the spatial dependence of the concentration of particles in the region where there is only diffusion current (that is, in the region $z \geq 0$), and in stationary regime, that is, in a regime where there is no temporal dependence.

We can use Equation (1.103) to find the spatial dependence of the concentration of holes. First, notice that the temporal partial derivative on the right-hand side of Equation (1.103) is null due to the assumption of stationary regime. Thus, for stationary regime, Equation (1.103) becomes:

$$-\frac{\partial J}{\partial z} = |q|\frac{p}{\tau_p} \tag{1.105}$$

We also know by assumption that we have only diffusion current for $z \geq 0$. On the other hand, with the help of Equation (1.90), we have:

$$J = J_{dif} = -D_p\frac{\partial \rho}{\partial z} = -|q|D_p\frac{\partial p}{\partial z}$$

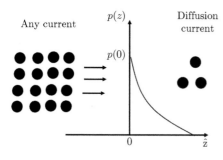

Any current $p(z)$ Diffusion current

$p(0)$

0    $\hat{z}$

Figure 1.11 Behaviour of the concentration of particles $p(z)$ when transport is through diffusion. In this example, the transport is solely through diffusion in the region of positive z. Consequently, there is an exponential decay of the concentration for $z \geq 0$.

$$\textit{for } z \geq 0 \qquad (1.106)$$

The subindex p in the diffusion coefficient reminds us that this coefficient may be different for electrons and holes. Using Equation (1.106) in Equation (1.105) gives:

$$\frac{\partial^2 p}{\partial z^2} = \frac{p}{D_p \tau_p} \qquad (1.107)$$

The solution to this differential equation is:

$$p(z) = A \cdot \exp\left(-\frac{z}{L_P}\right) + B \cdot \exp\left(+\frac{z}{L_P}\right) \qquad (1.108)$$

Where L_P is the diffusion length, given by:

$$L_P = \sqrt{D_p \tau_p} \qquad (1.109)$$

Now we apply boundary conditions to Equation (1.108). Recall that we are interested in the region $z \geq 0$. Since there is a loss mechanism, it makes sense that we require that the concentration goes to zero as z goes to infinity. This requirement sets $B = 0$. It also makes sense to express the result in terms of $p(0)$, which is the concentration at the origin. One can easily identify that $A = p(0)$, thus Equation (1.108) is reduced to:

$$p(z) = p(0) \cdot \exp\left(-\frac{z}{L_P}\right) \qquad (1.110)$$

It is a good exercise to repeat this process for negative charges. If you do that (and I encourage you to do it), you will find the same expression, but now in terms of n:

$$n(z) = n(0) \cdot \exp\left(-\frac{z}{L_N}\right) \qquad (1.111)$$

Thus, we have reached the important conclusion that, **when transport is by diffusion, then the concentration of particles decay exponentially in space**. This is why L_P and L_N are called diffusion lengths: they are, respectively, the typical distances that holes and electrons diffusive before they recombine. These important expressions will be used in the next chapters.

1.12 Suggestions for further reading

In this chapter, I tried, as far as possible, to keep the notation and logical flow conversant with Charles Kittel: *Thermal Physics*. This is a standard book, found in most libraries, and it is adequate for students seeking to pursue further studies in statistical physics. As another nice introductory book on statistical physics, with good qualitative discussions, I also recommend Ashley Carter: *Classical and Statistical Thermodynamics*. For more advanced students, the classic reference is Reif: *Fundamentals of Statistical and Thermal Physics*.

Concerning quantum mechanics, for the topics covered thus far, I recommend the first chapter of Cohen-Tannoudji: *Quantum Mechanics*, Vol. 1. Both volumes 1 and 2 are excellent, so I also recommend them to students seeking a solid foundation in quantum mechanics. This classic book is neatly divided into basic and advanced topics, so it is adequate for both beginners and more advanced students. Another excellent, and arguably more accessible, book on quantum mechanics is David Griffiths: *Introduction to Quantum Mechanics*.

1.13 Exercises

Exercise 1

(a) A system consists of a single electron trapped in a one-dimensional quantum well of length L. Using the requirement that the phase accumulation in a round trip must be a multiple of 2π, find the possible energy levels of the electrons.

(b) Suppose that the electron is in the ground state (which is the state with the lowest possible energy). If light is shone onto the quantum well, the electron can jump to states with higher energy upon absorption of a photon. The energy of each photon is proportional to the frequency of the radiation, and the proportionality is given by Planck's constant. Using energy conservation, find an expression for the wavelength of the photon that is required to excite the electron from the ground state to the second state (the state with the second lowest energy). Repeat the exercise for the transition between the ground state and the third state.

(c) Now use the results of item b) to find the lengths $L's$ of the quantum well that results in transition for red, green and blue photons (look up the wavelength range for each colour and pick one for the calculation).

Exercise 2

A system consists of a single electron in a three-dimensional cubic quantum well. Suppose that the side of the cube is L. If $g(E)$ is the multiplicity for the energy E, find:

(a) $g\left(\dfrac{3h^2}{8L^2 m}\right)$

(b) $g\left(\dfrac{6h^2}{8L^2 m}\right)$

Exercise 3
A system is in thermal equilibrium with a reservoir at temperature T. Derive an expression for the probability of the system to be in a given state with energy U.

Exercise 4
Consider the following set of true affirmations:

(a) At thermal equilibrium, the entropy of a system is maximized.

(b) The first derivative of a function at a point of maximum is zero.

(c) The temperature is the inverse of the derivative of the entropy.

The affirmations a)–c) are all true. Now consider the following conclusions:

(A) From a) and b), it follows that the derivative of the entropy of a system at thermal equilibrium is zero.

(B) From A) and c), it follows that the temperature of a system in equilibrium is infinite.

Conclusion B) is obviously fallacious. But why? Where is the logical flaw: in A) and/or B)?

Exercise 5
According to the fundamental assumption of statistical physics, all states are equally probable. Thus, if there are N states in a system, the probability of finding a given state must be $1/N$. However, according to Equation (1.24), the probability is not $1/N$, but

$$P(\beta) = \frac{g_B(U_0 - \varepsilon_\beta)}{N}$$

Resolve the apparent paradox.

Exercise 6
We have seen that the average energy of an ideal gas is $\langle U \rangle = (3/2)k_B T$ (Equation (1.62)). Thus, if there are N particles in the gas, the total energy

must be $U = \langle U \rangle N = (3/2) N k_B T$. Find an expression for the multiplicity of an ideal gas as a function of the energy U and the number of particles N.

Exercise 7

Suppose you want to modify Equation (1.24) to give the probability of a certain energy, instead of certain state, to be observed. What do you need to do with the numerator of the equation?

Exercise 8

Temperature is often the bogeyman of semiconductor industry: materials or effects that work wonderfully well at low temperatures go to the drain at room temperature. Suppose you are an engineer trying to make an absorber out of the system "one electron in a box". For your system to work, your one electron must stay in the ground state. Using the results of Exercise 1c, find the probability that the electron is in the ground state at room temperature. So, do you think it is going to work, or do you think that the electron will be excited to other states due to thermal energy?

Exercise 9

Two systems, A and B, are in thermal equilibrium (same temperature), but not in diffusive equilibrium. The two systems are put in diffusive contact, which means that they can exchange particles. Show that the total entropy increases when particles flow from the system with higher chemical potential to the system with lower chemical potential.

Exercise 10

The chemical potential of an ideal gas is given by:

$$\mu = k_B T \ln \left(\frac{n}{n_Q} \right)$$

where n is the concentration and $n_Q = \left(\dfrac{2 \cdot \pi \cdot m \cdot k_B \cdot T}{h^2} \right)^{3/2}$.

You do not need to worry about how to derive this equation yet, but you should be able to derive it after studying Chapter 2.

If gravity plays a role, then the total potential is:

$$\mu_{total} = k_B T \ln \left(\frac{n}{n_Q} \right) + mgy$$

where y is the height of the gas.

Find an expression giving the dependence of the concentration on the height for a gas in equilibrium. Your result is a fair approximation to the concentration of air in the atmosphere.

Exercise 11
Suppose that systems A and B in the figure below have different chemical potentials for free electrons. The systems are initially neutral, that is, they have zero net charge. If they are allowed to interact, then electrons diffusive from system B to system A, leading to charge accumulation and the setting up of an electric field. Suppose that, at equilibrium, the electric field is uniform and concentrated in the region d as shown in the figure. Find the magnitude and direction of the electric field (the magnitude must be given in terms of the chemical potential difference).

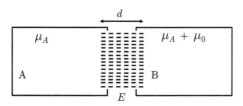

Exercise 12
When a ball is dropped from a certain height and gains velocity, we say that gravitational potential energy has been transformed into kinetic energy. Analogously, when charges flow through a resistor, we say that electrostatic potential energy is being transformed into heat. In both cases, there is a force driving the bodies.

Now, consider once again the problem of Exercise 11. Before equilibrium is reached, there is a diffusion current flowing, which means that there is energy dissipation, similarly to the energy dissipation in a resistor. But now there is no force involved. So, where is the energy coming from?

Exercise 13
Starting from the Gibbs sum, derive the Fermi–Dirac distribution and explain its physical meaning. Look up the Fermi–Dirac distribution as a function of energy for different temperatures and compare the curves. What is happening with higher energies as the temperature is increasing? What is happening with lower energies? Why?

Exercise 14
Consider the system "one electron trapped in a quantum well". The system is in thermal contact with a reservoir at temperature T. We learned that the probability of finding this system in the state $c_x = 1$, $c_y = 1$, $c_z = 1$ is given by Equation (1.31). But, since there is only one electron in the system, the probability of finding the system in the state $c_x = 1$, $c_y = 1$, $c_z = 1$ is logically equivalent to the probability of finding the electron in the orbital $c_x = 1$, $c_y = 1$,

$c_z = 1$. Thus, why is this probability given by Equation (1.31) instead of the Fermi–Dirac distribution?

Exercise 15

Consider a system A comprising the four orbitals O_1, O_2, O_3 and O_4. Orbitals O_1 and O_2 have the same energy ε_1, whereas orbitals O_3 e O_4 have the same energy ε_2. System A is in thermal contact (but no diffusive) with a reservoir at temperature T. If system A has only two electrons, what is the probability of orbital O_1 to be occupied?

Exercise 16

Suppose that the mean energy of an ideal gas of electrons is $\langle U \rangle$, and its chemical potential is μ. Express the probability of finding a given orbital of this gas occupied as function of $\langle U \rangle$, μ and the energy of the orbital.

Exercise 17

Explain how Equation (1.31) could be used to find the most probable energy of a system A in thermal contact with a reservoir. If system A is microscopic, can we affirm that it is virtually certain that this most probable property is the actual property of the system? Could we affirm it if system A was macroscopic?

2

Semiconductors

Learning objectives

In this chapter you will learn what a semiconductor is and the two types of charge carriers that contribute to its current: free electrons and holes. Then we will apply the concepts learned in the previous chapter to find expressions for the concentrations of these charges and relate them to the Fermi level. Finally, you will learn that it is very easy to change the electrical properties of semiconductors by means of a process known as doping. We are particularly interested in knowing how doping affects the charge carrier concentrations and the Fermi level.

In the previous chapter we covered the essential pre-requisites to study the physics of semiconductor devices. We are now in a good position to apply those concepts. We begin by defining what a semiconductor is, which requires a brief introduction to the important band theory

2.1 Band theory

In the previous chapter, we found the states of electrons trapped in a quantum well, which eventually we denoted by the term "orbitals". Because the quantum well is a very simple system, we managed to find these orbitals without delving into quantum mechanics. If we wanted to find the orbitals of more complex systems, such as atoms, we would need to solve the famous Schrödinger equation to

Essentials of Semiconductor Device Physics, First Edition. Emiliano R. Martins.
© 2022 John Wiley & Sons Ltd. Published 2022 by John Wiley & Sons Ltd.
Companion website: www.wiley.com/go/martins/essentialsofsemiconductordevicephysics

find the wavefunctions of the electrons. Though the mathematics would be far more complex, the essence of the physics is still the same: the states are discrete due to the confinement of electrons. In the case of atoms, the confinement is due to the electrostatic attraction between the nucleus and the electrons. The procedure to find the atomic orbitals can be found in any book on quantum mechanics (see section 1.12 for bibliographical suggestions) and we will not pursuit it here, but you may recognize from chemistry courses that these atomic orbitals are labelled by the letter *s, p, d, f*.... Our job in this section is to describe qualitatively what happens to the atomic orbitals as atoms are allowed to interact to form a solid.

To better illustrate what happens to the atomic orbitals as isolated atoms are allowed to interact, let us use an analogous mechanical system. For the sake of argument, suppose that we begin with a hydrogen atom, which has a single electron. An analogous mechanical system is the harmonic oscillator – the spring-mass system, as illustrated in Figure 2.1. Indeed, as the electron is trapped by the nucleus, so is the mass trapped by the spring. Furthermore, as the electron has well-defined energy states, so has the spring-mass system a well-defined resonance frequency (here the analogy breaks a little, because whereas there are many energies that the electron can access – as there are many orbitals, there is only one resonance frequency for the spring-mass system). Thus, the question of what happens to the atomic orbitals when two atoms interact is analogous to the question of what happens when two harmonic oscillators couple together. To gain intuition about the problem, we will briefly review what happens to the latter system, and then qualitatively extend the main result to the atomic orbital.

Consider a harmonic oscillator whose rest point (where the force is null) is at $x = a$, as represented in Figure 2.1.

According to Hooke's law, the force applied by the spring is proportional to the offset from the rest point. Thus, assuming a spring constant k_0 and mass m, Newton's second law for the mass m reads:

$$\frac{d^2 x}{dt^2} + \frac{k_0}{m}(x - a) = 0 \qquad (2.1)$$

The solution to this equation is a cosine function:

$$x = A \cdot \cos{(\omega_0 t + \varphi)} \qquad (2.2)$$

The phase and amplitude of the cosine function are determined by the initial conditions. We are interested,

Figure 2.1 A harmonic oscillator. The rest point, where the force is null, is at $x = a$.

however, in the resonance frequency ω_0, which is given by:

$$\omega_0 = \sqrt{\frac{k_0}{m}} \qquad (2.3)$$

I am assuming that you are already familiar with this story. If, however, you are not, then plug Equation (2.2) back into Equation (2.1) and check that it indeed satisfies the differential equation for ω_0 as given by Equation (2.3).

The harmonic oscillator of Figure 2.1 is analogous to an isolated atom, and the resonance frequency is analogous to the energy of the atomic orbital of the isolated atom. When two atoms interact, then we have a system that is analogous to two coupled harmonic oscillators, as illustrated in Figure 2.2. The coupling is achieved by the red spring, whose constant is Ω. Notice that, for simplicity's sake, I am considering two identical oscillators (same mass and same spring).

The mathematical description of the coupled harmonic oscillator is straightforward: one just applies Newton's law to both masses, which leads to the following two coupled differential equations:

$$\frac{d^2 x_1}{dt^2} + \frac{k_0}{m}(x_1 - a) + \frac{\Omega}{m}(x_1 - x_2 - b) = 0 \quad (1)$$

$$\frac{d^2 x_2}{dt^2} + \frac{k_0}{m}(x_2 + a) - \frac{\Omega}{m}(x_1 - x_2 - b) = 0 \quad (2) \qquad (2.4)$$

Equation (2.4)(1) is Newton's law applied to the mass on the right, whose coordinate is x_1, while Equation (2.4)(2) is Newton's law applied to the mass on the left, whose coordinate is x_2. Notice that b is the rest length of the coupling spring (the coupling force is null when $x_1 - x_2 = b$).

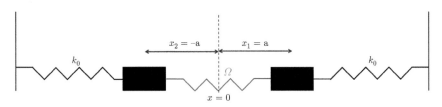

Figure 2.2 Two coupled harmonic oscillators. The spring in red, with constant Ω, represents the coupling.

To clean up the notation, we define $\Upsilon = \sqrt{\Omega/m}$ and rewrite Equation (2.4) as:

$$\frac{d^2 x_1}{dt^2} + \omega_0^2(x_1 - a) + \Upsilon^2(x_1 - x_2 - b) = 0 \quad (1)$$

$$\frac{d^2 x_2}{dt^2} + \omega_0^2(x_2 + a) - \Upsilon^2(x_1 - x_2 - b) = 0 \quad (2)$$

$$(2.5)$$

where ω_0 is given by Equation (2.3).

Because Equation (2.5)(1) and Equation (2.5)(2) are coupled (meaning that the former depends on x_2 and the latter on x_1), their solutions differ from the uncoupled system. The general method to find the solution of coupled equations is to diagonalize them, which involves finding their eigenvalues and eigenvectors. For our purpose, it is easier to uncouple them straight away by inspection. First, notice that we can find an equation that depends only on $x_1 + x_2$ if we add both equations together, thus getting:

$$\frac{d^2(x_1 + x_2)}{dt^2} + \omega_0^2(x_1 + x_2) = 0 \quad (2.6)$$

Since the masses are equal, the centre of mass is:

$$x_{cm} = \frac{x_1 + x_2}{2} \quad (2.7)$$

Thus:

$$\frac{d^2 x_{cm}}{dt^2} + \omega_0^2 x_{cm} = 0 \quad (2.8)$$

According to Equation (2.8), the centre of mass of the system behaves as a single (and hence uncoupled) harmonic oscillator, with mass m and spring constant k_0. Thus, the resonance frequency of the uncoupled oscillator ω_0 is still a resonance frequency of the coupled system for the oscillation of the centre of mass.

We can also get an equation involving a single variable by subtracting Equation (2.5)(2) from Equation (2.5)(1), thus getting:

$$\frac{d^2(x_1 - x_2)}{dt^2} + \omega_0^2(x_1 - x_2) + 2\Upsilon^2(x_1 - x_2) - 2(\omega_0^2 a + \Upsilon^2 b) = 0 \quad (2.9)$$

Denoting the relative distance between the masses as x_R:

$$x_R = x_1 - x_2 \quad (2.10)$$

we get:

$$\frac{d^2 x_R}{dt^2} + \left(\omega_0^2 + 2\Upsilon^2\right) x_R - 2\left(\omega_0^2 a + \Upsilon^2 b\right) = 0 \qquad (2.11)$$

The last term of Equation (2.11) is just a constant, so this is also the equation of an uncoupled harmonic oscillator, whose resonance frequency is:

$$\omega_R = \sqrt{\omega_0^2 + 2\Upsilon^2} \qquad (2.12)$$

and whose solution is:

$$x_R = A \cdot \cos\left(\omega_R t + \phi\right) + \frac{2\left(\omega_0^2 a + \Upsilon^2 b\right)}{\omega_R^2} \qquad (2.13)$$

Thus, there are two possible modes of oscillation: one at frequency ω_0 and the other at frequency ω_R.

The interpretation of this effect is the following: suppose we begin with two uncoupled harmonic oscillators (that is, with $\Upsilon = \Omega = 0$). We then have two modes of oscillations (each mass can oscillate on its own), but they have the same resonance - indeed, according to Equation (2.12), if $\Upsilon = 0$, then $\omega_R = \omega_0$. When the masses are coupled together, there are still two modes of oscillation and two resonance frequencies, but now they are no longer the same: one is still ω_0, but the other is $\omega_R \neq \omega_0$. Thus, we conclude that the effect of coupling is the "splitting of the resonance frequencies": they are no longer the same. The technical way of describing this is to say that "coupling breaks the degeneracy of the system". This term "degeneracy" comes from linear algebra and signifies that two or more eigenvectors have the same eigenvalue – here the resonance frequencies are the eigenvalues. Thus, a degeneracy is broken when they no longer have the same eigenvalue.

Notice that if there were three harmonic oscillators, then there would be three modes of oscillation and three resonance frequencies. Likewise, if there were n harmonic oscillators, then there would be n modes of oscillation and n resonance frequencies.

Analogously, if we consider two identical atoms sufficiently separated so that they do not interact, then the system has twice as many orbitals than in a single atom (since we have two atoms, the number of orbitals double), but the energies of two corresponding orbitals are the same. However, if these two atoms are allowed to interact, then the energy levels separate (they cease to be identical), because the orbitals themselves change due to the interaction. We thus speak of an "energy level splitting" due to the interaction (see Figure 2.3 for an illustration). If three atoms are allowed to interact, then each level splits into three new levels, and so on.

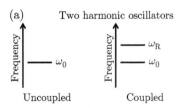

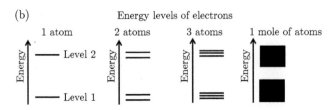

Figure 2.3 a) Illustration of resonance frequency splitting in a coupled harmonic oscillator. b) Analogous effect in the atomic energy levels when two or more atoms interact. When a macroscopic number of atoms interact to form a solid, there are so many energy levels, so close to each other, that we can effectively treat them as belonging to a continuum of energy levels. These ranges of energies with a continuum are called "energy bands". Notice that there may be a gap region between two bands: the energies falling in the gap are not allowed by the system.

What happens when many, many electrons are allowed to interact? If that happens, then there will be many, many energy levels very close together. Specifically, if a macroscopic number of electrons are allowed to interact, thus forming a solid, the levels are so close together that we can speak of an "energy band", that is, a region of energy where there are so many energy levels, and they are so close together, that they can be treated as if they formed a continuum of energies (see Figure 2.3b). Notice that there may be an "energy gap" between two energy bands: energies belonging to the gap are not allowed by the system.

The electrical properties of a solid depend not only on the formation of the bands, but also on their occupation. For example, suppose that the isolated atom, from which the solid is formed, has only one electron, and this electron occupies the ground-state energy level, which in Figure 2.3b is the "Level 1". When two atoms are allowed to interact, then "Level 1" splits into two, as shown in Figure 2.3b. As each atom contributes with one electron, both levels arising from the splitting of "Level 1" will be occupied. Likewise, as in the example of Figure 2.3b, if the lower band is formed from the splitting of "Level 1" by bringing n atoms together, then there will be n levels in the lower band, and n electrons, which means that the lower band will be fully occupied. Thus, if this level is occupied in the single atom, the entire energy band will be occupied by electrons. "Level 2" of the isolated atom, on the other hand, is unoccupied in our example. Since the upper band is formed by the splitting of this level, and this level is unoccupied in the isolated atom, then the upper band is completely unoccupied.

We could have other physical situations. For example, suppose that, in the isolated atom, "Level 1" denotes two overlapping levels (for example, due to spin degeneracy), but that only one of these two overlapping levels are occupied in the single atom. Thus, when allowing two atoms to interact, "Level 1" splits into four levels (two from each atom). However, only two of these four levels will be occupied, with each atom contributing with one electron. Thus, when a band is formed by the interaction of n atoms, there will be $2n$ levels in the band, but only n electrons. Consequently, the band is only partially occupied in this example. There could also be cases where the bands overlap, creating new bands with partial or complete occupation.

These examples show that the bands may be fully occupied, completely empty or partially occupied. It is convenient to represent the occupation by colouring the bands. Thus, in the first example of Figure 2.4, the lower band is fully occupied, and the upper band is fully empty. What would happen if an electric field was applied to a solid having these bands? In a classical description, the electric field would accelerate the electrons, thus increasing their energies. But if the band is fully occupied, there is no energy level (no orbital) available for the electrons to be accelerated into; consequently, the electrons "do not respond" to the applied field. Since the electrons cannot be accelerated by the electric field, then no electric current can be set up by the field. Thus, as indicated in Figure 2.4, such a material is an insulator.

Now suppose that one of the bands is partially occupied, as shown in the second example of Figure 2.4. The electrons in the upper band can contribute

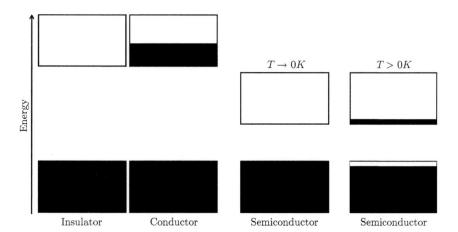

Figure 2.4 The electrical properties of the solids depend on the occupation of the energy bands.

to the current, because there are empty energy levels (orbitals) for them to be accelerated into. Thus, such a material is a conductor. The actual conductivity, however, depends on how easily the electrons can be moved by the field (which is a measure of their "mobility"), and of the concentration of electrons.

In the case of most interest to our purposes, the lower band is fully occupied, and the upper band is fully empty, as illustrated by the third example of Figure 2.4. This is the defining characteristic of a semiconductor, but it is only perfectly observed at the temperature of 0 K. Indeed, the energy difference between the bands of semiconductors is typically sufficiently low to allow a few electrons from the lower band to be thermally excited to the upper band, as represented in the fourth example of Figure 2.4. The electrons that are excited to the upper band can now contribute to the current, so the upper band is called the "conduction band". Moreover, these electrons interact weakly with the atomic structure, and thus behave as if they were trapped in an empty box, like in the quantum well system of the previous chapter. For these reasons, the electrons in the conduction band are called "free electrons". These are the electrons that contribute to the current.

The lower band in a semiconductor is called the valence band, and the energy difference between them is called the "band gap", but we will come back to these definitions in more detail in section 2.3. For now, we need to investigate what happens to the valence band when electrons jump to the conduction band, leaving "holes" behind.

2.2 Electrons and holes

As mentioned in the previous section, a semiconductor is a solid characterized by two energy bands: the valence band and the conduction band. Furthermore, the defining characteristic of a semiconductor is that, at 0 K, the valence band is fully occupied, and the conduction band is completely empty. Such condition, however, is only perfectly satisfied at 0 K. Indeed, thermal energy can excite electrons from the valence band to the conduction band, thus leaving behind a "hole". The purpose of this section is to explain how these holes left in the valence band act as charge carriers, similarly to the free electrons (that is, the electrons in the conduction band).

Hole theory is an important and complex part of semiconductor physics. Fortunately, for our purposes, a brief qualitative description suffices. One caveat, however, may be apposite: we are going to use a classical approach, thus treating the electrons essentially as particles. A full description, of course, requires quantum mechanics, which is outside the scope of this book.

Suppose that we have a solid formed by an atom with four valence electrons, as illustrated in Figure 2.5. The atom is locally and globally neutral. By being globally neutral, one means that the net charge in the solid is zero. By being locally neutral, one means that the net charge in a region containing a few atoms is also zero. But now consider what happens when an electron in the valence band is excited to the conduction band: this electron is now free to "move around" the solid, thus leaving the region around the vacant orbital bereft of a negative charge. The electron did not jump out of the solid, of course, so the global net charge is still zero. But its absence created a local positive charge. Moreover, this positive charge "occupies" the orbital left vacant by the electron. If an electric field is applied to this solid, a neighbouring valence electron can pick up

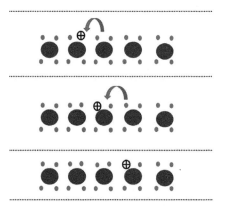

Figure 2.5 Illustration of hole transport. Each atom has four valence electrons. When one electron is excited to the conduction band, it becomes free to roam the semiconductor. Thus, the absence of the electron creates a local positive charge near the atom that has been vacated. When an electric field is applied to the solid, a neighbouring electron can pick up enough energy to jump to the empty orbital, thus leaving its own orbital empty. The net effect is that the local positive charge moves in the opposite direction of the electron. This movement sets up an electric current of holes. Holes can be treated as real particles, carrying not only charge but also momentum.

enough energy to jump from its orbital to the previously empty orbital. As illustrated in Figure 2.5, when this electron jumps to the empty orbital, it leaves its own orbital unoccupied, and its region with a net positive charge. If the electron goes to the left, then the positive charge goes to the right. Thus, a current can be set up by the movement of these positively charged "holes", which were left behind by the electrons when they were excited to the conduction band.

Though the holes are an absence of electrons, they can be treated as real particles, with momentum and charge: its momentum is essentially the momentum of the electrons in the valence band, and its charge has the same magnitude of the charge of electrons, but opposite sign, of course.

Thus, we conclude that there are two types of conduction in a semiconductor: conduction by the negatively charged electrons in the conduction band, which are called "free electrons", and conduction by the positively charged holes in the valence band. These two types of conduction are essential to the operation of semiconductor devices, as will be described in the next chapter.

As the actual conductivity of the semiconductor depends on the concentration of the charge carriers, we need to take a deeper look at what these concentrations depend on, which is the subject of the next section.

2.3   Concentration of free electrons

The electrical properties of any material, including semiconductors, depend on the concentration of charge carriers. In the previous section, we saw that there are two types of charge carriers in a semiconductor: free electrons and holes. In this section, we apply the concepts learned in Chapter 1 to find expressions for the concentrations of charge carriers. We focus attention on the procedure for free electrons, and in a later section we extend it to find the concentration of holes. Fortunately, the logic and calculations for free electrons and holes are very similar.

The concentration of free electrons is, by definition, the ratio of the number of electrons in the conduction band to the volume of the semiconductor. So, our first job is to calculate the number of electrons in the conduction band.

Before we begin, let us remind ourselves of the physical context. The semiconductor is a macroscopic system in thermal contact with a reservoir at a temperature T. If we look at a specific orbital of the conduction band, can we know whether it is occupied or not? No, we cannot, the system is too complicated to determine that. But, as we saw in Chapter 1, though we cannot know for sure if the orbital is occupied, we do know the probability of it being occupied: the probability is given by our cherished Fermi–Dirac distribution.

Now consider the following example: suppose there is a funny classroom in magical semiconductor land. You are told that the probability of each seat being occupied by a student in the funny classroom is 0.5 (we are taking the probability to range from 0 to 1, not in percentage). Suppose that there are only two seats in this classroom. How many students are in the classroom? Of course, you cannot answer this question with certainty, as you only know the probability. So, it may be that there will be no student, or one student or two students. You know that the probability of finding no student is 0.25, of finding one student is 0.5, and of finding two students is 0.25. So, the best guess is that there is only one student, but this is not a great guess, since it is as likely as not that it will turn out to be correct. Now, suppose that there are, say, four million seats in this funny classroom. In this case, it is a pretty good guess to assert that there are two million students in the classroom. In fact, the probability of finding two million students is so close to 1, within such an insignificant margin of error (this margin is the "fractional width" discussed in Chapter 1), that the answer can no longer be called a guess. Thus, if the system is large, then we can find a virtually exact answer by just adding the probability of each seat being occupied (in this example, by adding 0.5 four million times).

The idea I want to convey with this example is that, if we are dealing with a macroscopic system, then we can find the number of electrons with virtual certainty by just adding up the probability of finding each orbital occupied. Again, we are taking advantage of the fact that statistics becomes virtually an exact science in large systems. Thus, since the semiconductor is a macroscopic system, we can obtain the number of electrons in the conduction band, which we denote by N_e, as:

$$N_e \cong \sum_{cb} f(\varepsilon_{cb}) = \sum_{cb} \frac{1}{1 + \exp\left(\dfrac{\varepsilon_{cb} - \mu}{k_B T}\right)} \tag{2.14}$$

where $f(\varepsilon_{cb})$ is the Fermi–Dirac distribution, and Equation (1.66) has been used.

Notice that the subscript cb has been added to the energy ε_{cb} to emphasize that we are considering only orbitals that belong to the conduction band. Furthermore, the subscript is also used to indicate that the sum in Equation (2.14) is carried out for the orbitals in the conduction band.

Before we focus attention on how to compute the sum of Equation (2.14), it is convenient to adapt the notation to the literature on semiconductors. Thus, as discussed in section 1.7, the potential μ is called the "Fermi level", a term which we adopt from now on and denote it by ε_F. Using the new notation, Equation (2.14) reads:

$$N_e = \sum_{cb} f(\varepsilon_{cb}) = \sum_{cb} \frac{1}{1 + \exp\left(\dfrac{\varepsilon_{cb} - \varepsilon_F}{k_B T}\right)} \tag{2.15}$$

It is also convenient to ascribe a notation to highest energy of the valence band, the lowest energy of the conduction band, and the energy difference between them. As shown in Figure 2.6, these energies are respectively denoted by ε_V, ε_C and ε_G, where:

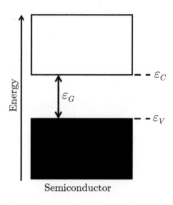

$$\varepsilon_G = \varepsilon_C - \varepsilon_V \qquad (2.16)$$

Thus, ε_G is the "band gap".

Figure 2.6 Definitions of ε_V, ε_C and ε_G.

Now we focus our attention on how to compute N_e. This is such a crucial calculation in the field of semiconductors, that we are going to do it twice: in this, and in the next section.

At first glance, the computation of N_e appears to be a herculean job. Indeed, before we even begin doing the sum, we need to find the orbitals of the conduction band, and their energies. Fortunately, this is not such a difficult task, because we have already found these orbitals. Recall that the electrons in the conduction band are "free", that is, they interact weakly with the atomic structure. Thus, the system "free electrons in a semiconductor" is very similar to the system "electrons in a box", for which we found the orbitals and energies in Chapter 1. In the literature, this system is often called an "electron gas". There are, however, two modifications that need to be made, but they are quite straightforward.

The first modification has to do with the interaction with the atomic structure: though it is weak, it is not non-existent. But the interaction can be straightforwardly accounted for by using an "effective mass" for the electrons, which is denoted by m^*. The idea of incorporating the interaction with the atomic structure through an effective mass is analogous to the idea of treating the physics of a body inside a swimming pool by ascribing an effective mass to the body. To be fair, the concept of effective mass is far more interesting than that, but we cannot delve into it here; so, we will take it as a given parameter. However, I cannot help mentioning one quite interesting feature: because electrons have wave properties, the effective mass depends on interference effects; in some cases, destructive interference can lead to a negative effective mass, which means that the electron is accelerated in the opposite direction of the force!!! Electrons are funny creatures indeed.

The second modification takes into account the fact that the energies of the system "free electrons in a semiconductor" have an energy offset of ε_C in comparison with the system "electrons in a box". Physically, the offset is associated with the potential energy that prevents electrons from flying off the semiconductor. This offset can also be easily accounted for by adding the term ε_C to Equation (1.58).

Thus, applying both modifications to Equation (1.58), it becomes:

$$\varepsilon_{cb} = \frac{h^2}{8L^2 m_e^*}\left(c_x^2 + c_y^2 + c_z^2\right) + \varepsilon_C \tag{2.17}$$

where ε_{cb} is the energy of the orbital (c_x, c_y, c_z) of the conduction band, m_e^* is the electron effective mass and L is the length of the semiconductor (recall that, for simplicity, we are treating the box, that is, the semiconductor, as a cube; as it will be seen shortly, this simplification does not affect the calculation of the concentration).

Thus far, we have blatantly ignored the effect of spin, but we cannot get away with it any longer. Fortunately, the inclusion of spin is trivial: it only means that each triplet (c_x, c_y, c_z) denotes two orbitals, one with spin up and another with spin down. So, to include the effect of spin, we multiply the sum over all (c_x, c_y, c_z) by two. Thus, with the help of Equation (2.17), the sum in Equation (2.15) becomes:

$$N_e = \sum_{cb} \frac{1}{1 + \exp\left(\dfrac{\varepsilon_{cb} - \varepsilon_F}{k_B T}\right)}$$

$$= 2 \sum_{c_x = 1}^{\infty} \sum_{c_y = 1}^{\infty} \sum_{c_z = 1}^{\infty} \frac{1}{1 + \exp\left[\dfrac{\left(\dfrac{h^2}{8L^2 m_e^*}\left(c_x^2 + c_y^2 + c_z^2\right) + \varepsilon_C\right) - \varepsilon_F}{k_B T}\right]} \tag{2.18}$$

Typically, the energy of an electron in the conduction band is much higher than the "thermal energy" $k_B T$. Thus, it is often true that

$$\exp\left[\dfrac{\left(\dfrac{h^2}{8L^2 m_e^*}\left(c_x^2 + c_y^2 + c_z^2\right) + \varepsilon_C\right) - \varepsilon_F}{k_B T}\right] \gg 1$$

As an example of the values taken by the exponential, consider silicon at room temperature. The band gap of silicon is about 1.14 $e.V$ ($electron-volts$). The thermal energy at room temperature ($T = 300\ K$), on the other hand, is

$k_B T = 0.0258$ e. V. That means that the value of the exponential is 2326 for a separation of only 0.2 e. V between the energy in the conduction band and the Fermi level. As we will prove later in this chapter, the Fermi level lies in the middle of the band gap, so the separation for silicon is at least 0.57 e. V.

We thus take advantage of the large number of the exponential to simplify the denominator in the sum of Equation (2.18) as:

$$1 + \exp\left[\frac{\left(\frac{h^2}{8L^2 m_e^*}\left(c_x^2 + c_y^2 + c_z^2\right) + \varepsilon_C\right) - \varepsilon_F}{k_B T}\right]$$

$$\cong \exp\left[\frac{\left(\frac{h^2}{8L^2 m_e^*}\left(c_x^2 + c_y^2 + c_z^2\right) + \varepsilon_C\right) - \varepsilon_F}{k_B T}\right] \tag{2.19}$$

Semiconductors that satisfy this condition are called "non-degenerate". Using the approximation for non-degenerate semiconductors we get:

$$N_e = 2\sum_{c_x}\sum_{c_y}\sum_{c_z} \frac{1}{\exp\left[\frac{\left(\frac{h^2}{8L^2 m_e^*}\left(c_x^2 + c_y^2 + c_z^2\right) + \varepsilon_C\right) - \varepsilon_F}{k_B T}\right]}$$

$$\therefore$$

$$N_e = 2\sum_{c_x}\sum_{c_y}\sum_{c_z} \exp\left[-\frac{\left(\frac{h^2}{8L^2 m_e^*}\left(c_x^2 + c_y^2 + c_z^2\right) + \varepsilon_C\right) - \varepsilon_F}{k_B T}\right] \tag{2.20}$$

Isolating the terms of the sum that do not depend on the sum indices c_x, c_y and c_z:

$$N_e = 2\exp\left(-\frac{\varepsilon_C - \varepsilon_F}{k_B T}\right)\sum_{c_x}\sum_{c_y}\sum_{c_z} \exp\left[-\frac{\frac{h^2}{8L^2 m_e^*}\left(c_x^2 + c_y^2 + c_z^2\right)}{k_B T}\right] \tag{2.21}$$

We have already encountered an identical sum in Equation (1.59). The sum is most easily solved by approximating it to an integral. Such an approximation is valid because the exponential varies slowly with respect to variations in c_x, c_y and c_z. Notice that approximating the sum to an integral is tantamount to approximating the Riemann sum to an integral (also notice that the steps $\Delta c_{x,\,y,\,z}$ are unity, since c_x, c_y and c_z are natural numbers). Thus:

$$\sum_{c_x}\sum_{c_y}\sum_{c_z}\exp\left[-\frac{\dfrac{h^2}{8L^2 m_e^*}\left(c_x^2 + c_y^2 + c_z^2\right)}{k_B T}\right]$$

$$\approx \int_0^\infty\int_0^\infty\int_0^\infty \exp\left[-\frac{\dfrac{h^2}{8L^2 m_e^*}\left(c_x^2 + c_y^2 + c_z^2\right)}{k_B T}\right] dc_x dc_y dc_z \qquad (2.22)$$

This integral is tabulated and the result is, of course, the same as in Equation (1.60):

$$\int_0^\infty\int_0^\infty\int_0^\infty \exp\left[-\frac{\dfrac{h^2}{8L^2 m_e^*}\left(c_x^2 + c_y^2 + c_z^2\right)}{k_B T}\right] dc_x dc_y dc_z = \frac{L^3}{\left(\dfrac{h^2}{2\pi m_e^* k_B T}\right)^{3/2}}$$

$$(2.23)$$

Thus, substituting Equation (2.23) into Equation (2.21), we find that the number of free electrons is:

$$N_e = 2\exp\left(-\frac{\varepsilon_C - \varepsilon_F}{k_B T}\right)\frac{L^3}{\left(\dfrac{h^2}{2\pi m_e^* k_B T}\right)^{3/2}} \qquad (2.24)$$

Notice that L^3 is the volume of the semiconductor. Thus, the concentration of free electrons, which we denote by n, is:

$$n = \frac{N_e}{L^3} = 2\exp\left(-\frac{\varepsilon_C - \varepsilon_F}{k_B T}\right)\frac{1}{\left(\dfrac{h^2}{2\pi m_e^* k_B T}\right)^{3/2}} = n_c \exp\left(-\frac{\varepsilon_C - \varepsilon_F}{k_B T}\right)$$

$$\therefore n = n_c \exp\left(-\frac{\varepsilon_C - \varepsilon_F}{k_B T}\right) \qquad (2.25)$$

where:

$$n_c = \frac{2}{\left(\dfrac{h^2}{2\pi m_e^* k_B T}\right)^{3/2}} \qquad (2.26)$$

Equation (2.25) is the main result of this section. Notice that the concentration of free electrons depends on n_c, which depends on the temperature and on the atomic structure (the latter through the effective mass). Importantly, the concentration depends on the difference between ε_C and ε_F. Remember that ε_F is the electrochemical potential, which here coincides with the chemical potential (the electrostatic potential is only relevant when two systems with different potentials interact, which is not the case here, but it will be the case in the next chapter). Thus, Equation (2.25) relates the chemical potential with the concentration. In Chapter 1, we noticed qualitatively that these two parameters are intimately related. Now we have found an explicit relation for them.

Before we move on, I want to derive Equation (2.25) again, but this time using the important concept of "density of states".

2.4   Density of states

In the previous section, we derived the concentration of free electrons (Equation (2.25)) by a direct method: we just added up the occupancy probability of all orbitals in the conduction band. This method has clear pedagogical benefits, but it bypasses an important concept: the density of states. In this section, I will show you another strategy to obtain Equation (2.25), which uses the density of states. It is instructive to compare both methods and to verify that they amount to the same calculation, just done in a different way.

We begin by highlighting the difference between the two methods. In the previous section, we found the number of electrons by adding up the probability of finding each orbital to be occupied. We learned in Chapter 1 that this probability is given by the Fermi–Dirac distribution, which depends on the energy of the orbital. That means that orbitals with the same energy have the same probability. Thus, instead of adding up the probabilities of all orbitals independently, we could have grouped the orbitals with the same energy, and then add up their probabilities in one stroke. That can be done by finding how many orbitals have a given energy, and then multiply the number of orbitals by the Fermi–Dirac distribution evaluated at the given energy. For example, say that there were three orbitals with energy ε_0; the contribution of these orbitals to the overall sum would be given by the equation $f(\varepsilon_0) + f(\varepsilon_0) + f(\varepsilon_0) = 3f(\varepsilon_0)$. In the approach of the previous section, we added the three terms explicitly, as on the left-hand side of the equation; now, in this section, we will group them together, as on the right-hand side of the equation. One advantage of this approach is that it allows the sum to be performed over energies instead of over the triplets c_x, c_y and c_z. Another major advantage is that the information about how many orbitals have a given energy is captured by one single function: the density of states.

The concept of density of states is not complicated. First, recall what a density is. Consider, for example, the mass density of a rope. The mass density of a rope is, by definition, the mass of the rope in a unit of length. But how can we calculate, or measure, the mass density of the rope? If the rope is uniform, then all we need to do is to pick any length whatsoever, determine its mass and divide it by the length. If the rope is non-uniform, however, then we need to divide it in sections sufficiently small so that the rope is uniform within each section, and then determine the mass in each section, and then divide the mass of each section by the length of the section, thus obtaining the mass density as a function of position.

The idea behind the density of states is basically the same, but instead of mass, we have number of states, and instead of small sections of a rope, we have small energy ranges. So, to obtain the density of states, all we need to do is to find the number of orbitals inside a small energy range. Notice that, if we were to follow our terminology strictly, we would need to call it "the density of orbitals", but as nobody calls it "density of orbitals", we stick to "density of states".

Let us now express this idea in mathematical language. We denote the number of orbitals between the energy ε and the energy $\varepsilon + \Delta\varepsilon$ by $N_{\varepsilon \to \varepsilon + \Delta\varepsilon}$. The density of states – which we denote by $D(\varepsilon)$ – is, by definition, the number of orbitals within a small energy range, divided by "the length of the range". Mathematically, this definition is expressed as:

$$D(\varepsilon) = \frac{N_{\varepsilon \to \varepsilon + \Delta\varepsilon}}{\Delta\varepsilon} \qquad (2.27)$$

We can use the density of states to group all orbitals that have the same energy together, and then find their contribution in a single stroke, and then sum over all energies to find the total number of electrons. Thus, Equation (2.15) can be recast as:

$$N_e = \sum_{cb} f(\varepsilon_{cb}) = \sum_{\varepsilon} f(\varepsilon) N_{\varepsilon \to \varepsilon + \Delta\varepsilon} = \sum_{\varepsilon} f(\varepsilon) D(\varepsilon) \Delta\varepsilon \qquad (2.28)$$

Though the interpretation of Equation (2.28) is straightforward, it is worth pondering at it a little bit to make sure that the logic is clear. Notice that the first equality is the same as in Equation (2.15) of the previous section. I left it there to emphasize that we are still doing the same procedure we did in the previous section: adding up the probabilities of occupation over all orbitals. The only difference now is that we are grouping orbitals with the same energy, as shown in the second equality of Equation (2.28). This second equality is the mathematical form of the command "find the number of orbitals within a small energy range and multiply it by the Fermi–Dirac distribution; then move on to the next small energy range and do the same; and so on." There is, of course, an approximation in this second equality, in the sense that we are disregarding the difference between ε and $\varepsilon + \Delta\varepsilon$ in the Fermi–Dirac distribution (we are treating all energies between ε and $\varepsilon + \Delta\varepsilon$ as if they had the same probability $f(\varepsilon)$). This is not a problem, though, because we will soon treat $\Delta\varepsilon$ as an infinitesimal quantity, thus turning the sum into an integral. But, to do that, we need to express the second equality in a way that $\Delta\varepsilon$ appears explicitly. That is easily done using the density of states, that is, using Equation (2.27), thus resulting in the third equality of Equation (2.28).

Thus, in the form of the last equality of Equation (2.28), the sum is ready to be transformed into an integral by letting $\Delta\varepsilon \to 0$. Thus:

$$N_e = \lim_{\Delta\varepsilon \to 0} \sum_{\varepsilon} f(\varepsilon) D(\varepsilon) \Delta\varepsilon = \int_{\varepsilon_C}^{\infty} f(\varepsilon) D(\varepsilon)\, d\varepsilon \qquad (2.29)$$

Notice that, since we are including only the orbitals in the conduction band, the lower limit of the integral is ε_C. The upper limit can be set at ∞, because the Fermi–Dirac distribution goes to zero as the energy goes to infinity.

The expression for N_e in terms of the density of states, as shown in Equation (2.29), is the most common form found in the literature. As mentioned previously, this form is convenient in that it involves only a one-dimensional integral, and it only depends on the density of states, on ε_C and on the Fermi level (through the Fermi–Dirac distribution). Never forget, however, that this is just a sum over the occupancy probabilities in the conduction band.

Now we focus attention on how to compute the density of states. It is easier to begin by deriving the density of states of the system "electrons in a box", and then adapt it to the system "free electrons in a semiconductor". Furthermore, it is much easier to convey the main ideas in a two-dimensional system than in a three-dimensional one. So, we will first find the density of states of a two-dimensional box, and then apply the same concepts for a three-dimensional box.

The energy of a two-dimensional box (that is, the two-dimensional version of Equation (1.58)) is given by:

$$\varepsilon = \frac{h^2}{8L^2 m}\left(c_x^2 + c_y^2\right) \tag{2.30}$$

Notice that, in a two-dimensional system, each orbital is denoted by a pair (c_x, c_y) instead of the triplet of a three-dimensional system. To find $N_{\varepsilon \to \varepsilon + \Delta\varepsilon}$, we need to calculate how many orbitals lie in the energy range between ε and $\varepsilon + \Delta\varepsilon$. This amount to finding how many pairs (c_x, c_y) lie in the range between ε and $\varepsilon + \Delta\varepsilon$. To see how this can be done, it is instructive to visualize the distribution of these pairs in a two-dimensional plot, with axes given by c_x and c_y, as shown in Figure 2.7a. We represent each orbital by a dot in the corresponding coordinate. Notice that, even though c_x and c_y can take only natural numbers, for now we are allowing negative integers; later on, we will compensate for that. So, each dot in the graph of Figure 2.7a represents an orbital.

Our final goal is to find the number of orbitals in an energy range. The key insight is to notice that, since the energy is proportional to $c_x^2 + c_y^2$, then the energy of each orbital is defined by its distance to the origin. Denoting the distance as R, then:

$$R = \sqrt{c_x^2 + c_y^2} \tag{2.31}$$

Combining Equation (2.31) and Equation (2.30), we get:

$$R = \sqrt{\frac{8L^2 m}{h^2}}\sqrt{\varepsilon} \tag{2.32}$$

Since it is the distance to the origin that fixes the energy, then orbitals lying in a given energy range lie in a given ring of the graph, as shown in Figure 2.7b. We can then work out the number of orbitals lying inside a ring, and then use Equation (2.32) to find the number of orbitals in the corresponding energy range.

Adopting this strategy, we denote the number of orbitals lying inside a ring beginning in R and ending in $R + \Delta R$ by $N_{R \to R + \Delta R}$. To find $N_{R \to R + \Delta R}$, we can first calculate how many orbitals lie within a disc or radius R, which we denote by $\eta(R)$, as shown Figure 2.7c,d. Then, we can find $N_{R \to R + \Delta R}$ by the subtraction $\eta(R + \Delta R) - \eta(R)$.

The number of orbitals within the disc can be found straightforwardly by noting that the area of the graph of Figure 2.7 can be filled out by ascribing a

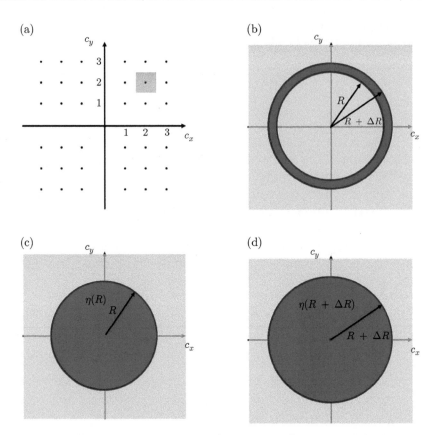

Figure 2.7 Two-dimensional representation of the orbitals of the system "electrons in a two-dimensional box". a) each orbital is denoted by a pair (c_x, c_y). Each pair is represented by a dot in the two-dimensional graph. Though c_x and c_y can take only natural numbers, we include negative numbers for now, and later compensate. b) We want to find how many orbitals lie in the energy range between ε and $\varepsilon + \Delta\varepsilon$. Since ε is proportional to $\left(c_x^2 + c_y^2\right)$, an energy range lies in a ring of the graph, so we need to find how many orbitals lie inside the ring. c) To find how many orbitals lie inside the ring, we can first find $\eta(R)$, which is the number of orbitals lying inside a disc of radius R. d) Then we can find $\eta(R + \Delta R)$; thus, the number of orbitals inside the ring of b) is found by the subtraction $\eta(R + \Delta R) - \eta(R)$.

square of side one to each dot. The square for the orbital $(2, 2)$ is shown in Figure 2.7a. The orbital is in the centre of the square of side one. Thus, the entire area can be filled out by placing these unit squares in each orbital. Now, consider the disc of radius R, as shown in Figure 2.7c. How many orbitals lie within the disc? If we ascribed a little square to each orbital, then the number of orbitals can be found by taking the ratio of the area of the disc (πR^2) to the area of each square (notice that this area is one, since the sides are one). We also need to multiply the result by two, to include the spin; but we also need to divide it by four, because we took the four quadrants into consideration, when only the quadrant with positive integers represents a physical orbital. Thus, we conclude that:

$$\eta(R) = \frac{\pi R^2}{1} \cdot \frac{2}{4} \tag{2.33}$$

Thus, the number of orbitals inside the ring of Figure 2.7b is given by:

$$N_{R \to R + \Delta R} = \eta(R + \Delta R) - \eta(R) = \frac{\pi (R + \Delta R)^2}{2} - \frac{\pi (R)^2}{2} \tag{2.34}$$

We could go on and compute Equation (2.34) explicitly. But, since what we actually want is $N_{\varepsilon \to \varepsilon + \Delta\varepsilon}$, we can already transform $\eta(R)$ into $\eta(\varepsilon)$, and then compute $N_{\varepsilon \to \varepsilon + \Delta\varepsilon}$ straight away. Indeed, since $\eta(\varepsilon)$ is the number of orbitals with energy smaller than ε, then, $N_{\varepsilon \to \varepsilon + \Delta\varepsilon}$ can be found straightforwardly as:

$$N_{\varepsilon \to \varepsilon + \Delta\varepsilon} = \eta(\varepsilon + \Delta\varepsilon) - \eta(\varepsilon) \tag{2.35}$$

To find $\eta(\varepsilon)$, we can substitute Equation (2.32) into Equation (2.33):

$$\eta(\varepsilon) = \frac{2}{4} \cdot \frac{\pi R^2}{1} = \frac{2}{4} \cdot \pi \frac{8L^2 m}{h^2} \varepsilon \tag{2.36}$$

We are almost there. From the definition of density of states (Equation (2.27)):

$$D(\varepsilon) = \frac{N_{\varepsilon \to \varepsilon + \Delta\varepsilon}}{\Delta\varepsilon}$$

Therefore:

$$D(\varepsilon) = \frac{N_{\varepsilon \to \varepsilon + \Delta\varepsilon}}{\Delta\varepsilon} = \frac{\eta(\varepsilon + \Delta\varepsilon) - \eta(\varepsilon)}{\Delta\varepsilon} \tag{2.37}$$

The density of states can now be found by taking the limit $\Delta\varepsilon \to 0$:

$$D(\varepsilon) = \lim_{\Delta\varepsilon \to 0} \frac{\eta(\varepsilon + \Delta\varepsilon) - \eta(\varepsilon)}{\Delta\varepsilon} = \frac{d\eta}{d\varepsilon} \tag{2.38}$$

With the help of Equation (2.36), we finally find:

$$D(\varepsilon) = \frac{2}{4} \cdot \pi \frac{8L^2 m}{h^2} = \frac{4\pi L^2 m}{h^2} \tag{2.39}$$

Equation (2.39) is the density of states of a two-dimensional electron gas.

We can follow the same reasoning to find the density of states of the three-dimensional system: the unit squares become unit cubes, the disc becomes a sphere, and instead of dividing by four we need to divide by eight to include only the part of the sphere with orbitals denoted by three positive integer numbers (there are three axis, so eight combinations of positives and negatives, out of which only one has three positive numbers). We still need to multiply by two because of the spin. Thus, the number of orbitals within a sphere of radius R is:

$$\eta_{3D}(R) = \frac{4\pi R^3}{3} \cdot \frac{2}{8} \tag{2.40}$$

Recall that $4\pi R^3/3$ is the volume of the sphere. In the three-dimensional case, the radius is:

$$R = \sqrt{c_x^2 + c_y^2 + c_z^2}$$

and the energy is:

$$\varepsilon = \frac{h^2}{8L^2 m}\left(c_x^2 + c_y^2 + c_z^2\right)$$

Thus:

$$\eta_{3D}(\varepsilon) = \frac{2}{8} \cdot \frac{4\pi R^3}{3} = \frac{2}{8} \cdot \frac{4\pi \left(\frac{8L^2 m}{h^2}\varepsilon\right)^{3/2}}{3} \tag{2.41}$$

Therefore, the density of states of the three-dimensional system is:

$$D_{3D}(\varepsilon) = \frac{d\eta_{3D}}{d\varepsilon} = \frac{\pi\left(\frac{8L^2 m}{h^2}\right)^{3/2}}{2}\sqrt{\varepsilon}$$

Rearranging:

$$D_{3D}(\varepsilon) = 4\pi L^3 \left(\frac{2m}{h^2}\right)^{3/2}\sqrt{\varepsilon} \tag{2.42}$$

Notice that the density of states is a function of the energy.

To complete this section, we need to adapt Equation (2.42) to the system "free electrons in a semiconductor". This again requires the same two modifications as before. Recall that one modification is to replace the mass with the effective mass, which we denote by m_e^*, where the subindex e

differentiates between electron and holes. The second modification takes into account the fact that the lowest energy is ε_C, which amounts to shifting the origin of the coordinate system of the orbitals from 0 to ε_C. Since the density of states depends on the distance to the origin, then all we need to do is to substitute $\varepsilon - \varepsilon_C$ for ε in Equation (2.42). Thus, denoting the density of states of free electrons in a semiconductor by $D_e(\varepsilon)$, and applying the two modifications to Equation (2.42), we get:

$$D_e(\varepsilon) = 4\pi L^3 \left(\frac{2m_e^*}{h^2}\right)^{3/2} \sqrt{\varepsilon - \varepsilon_C} \tag{2.43}$$

We can now find the number of free electrons by substituting Equation (2.43) into Equation (2.29). Using the approximation for non-degenerate semiconductors (Equation (2.19)) in the Fermi–Dirac distribution, we find:

$$N_e = \int_{\varepsilon_C}^{\infty} f(\varepsilon) D_e(\varepsilon)\, d\varepsilon = \int_{\varepsilon_C}^{\infty} \exp\left[-\frac{(\varepsilon - \varepsilon_F)}{k_B T}\right] D_e(\varepsilon)\, d\varepsilon$$

$$= \int_{\varepsilon_C}^{\infty} \exp\left[-\frac{(\varepsilon - \varepsilon_F)}{k_B T}\right] 4\pi L^3 \left(\frac{2m_e^*}{h^2}\right)^{3/2} \sqrt{\varepsilon - \varepsilon_C}\, d\varepsilon \tag{2.44}$$

Isolating the terms that do not depend on the energy:

$$N_e = 4\pi L^3 \left(\frac{2m_e^*}{h^2}\right)^{3/2} \exp\left(\frac{\varepsilon_F}{k_B T}\right) \int_{\varepsilon_C}^{\infty} \exp\left(-\frac{\varepsilon}{k_B T}\right) \sqrt{\varepsilon - \varepsilon_C}\, d\varepsilon \tag{2.45}$$

Integrating, we find:

$$N_e = L^3 \frac{2}{\left(\frac{h^2}{2\pi m_e^* k_B T}\right)^{3/2}} \exp\left(-\frac{\varepsilon_C - \varepsilon_F}{k_B T}\right) \tag{2.46}$$

Thus, the concentration of free electrons is:

$$n = \frac{N_e}{L^3} = n_c \exp\left(-\frac{\varepsilon_C - \varepsilon_F}{k_B T}\right) \tag{2.47}$$

where n_c is given by Equation (2.26).

Equation (2.47) is identical to Equation (2.25), as expected.

2.5 Concentration of holes and Fermi level

In the previous section we found an expression for the concentration of free electrons in a semiconductor using the concept of density of states. The same ideas can be applied to find the concentration of holes. Indeed, the expression for the density of states of holes is very similar to Equation (2.43). The main difference is that the reference energy is no longer ε_C, but ε_V. To be more specific, as we have seen in the previous section, the density of states of free electrons depends on $\sqrt{\varepsilon-\varepsilon_C}$. Recall that this difference is a measure of the distance between the orbital and the origin (the origin being fixed at ε_C). Analogously, the density of states of holes depends on the distance between the energy and ε_V, so the term $\sqrt{\varepsilon-\varepsilon_C}$ is replaced with the term $\sqrt{\varepsilon_V-\varepsilon}$. Notice that, while ε_C is the lowest energy in the conduction band, ε_V is the highest energy in the valence band; this feature explains why, inside the square root, we have $\varepsilon-\varepsilon_C$ for free electrons but $\varepsilon_V-\varepsilon$ for holes: in both cases the subtraction is always a positive number (as required by the notion of "distance"). Therefore, the density of holes, which we denote by $D_h(\varepsilon)$, can be straightforwardly found by replacing $\sqrt{\varepsilon-\varepsilon_C}$ with $\sqrt{\varepsilon_V-\varepsilon}$ in Equation (2.43). Accordingly:

$$D_h(\varepsilon) = 4\pi L^3 \left(\frac{2m_h^*}{h^2}\right)^{3/2} \sqrt{\varepsilon_V-\varepsilon} \qquad (2.48)$$

where m_h^* is the effective mass of holes.

To compute the number of holes, we also need to obtain the probability of finding an orbital "occupied by a hole". We denote this probability as $f_h(\varepsilon)$. An orbital that is occupied by a hole is an orbital that is unoccupied by an electron. Since the orbital must be either occupied or empty, then it follows immediately that $f(\varepsilon) + f_h(\varepsilon) = 1$ (recall that $f(\varepsilon)$ is the probability of finding the orbital occupied by an electron). Thus, the probability of finding an orbital in the valence band occupied by a hole is:

$$f_h(\varepsilon) = 1 - f(\varepsilon) \qquad (2.49)$$

Recalling that $f(\varepsilon)$ is the Fermi–Dirac distribution, we get:

$$f_h(\varepsilon) = 1 - \frac{1}{1+\exp\left(\dfrac{\varepsilon-\varepsilon_F}{k_B T}\right)} = \frac{1}{1+\exp\left(\dfrac{\varepsilon_F-\varepsilon}{k_B T}\right)} \qquad (2.50)$$

We can simplify this expression by using one more time the approximation for non-degenerate semiconductors:

$$f_h(\varepsilon) \approx \frac{1}{\exp\left(\dfrac{\varepsilon_F - \varepsilon}{k_B T}\right)} \qquad (2.51)$$

Thus, the total number of holes in the valence band is:

$$N_h = \int_{-\infty}^{\varepsilon_V} D_h(\varepsilon) f_h(\varepsilon)\, d\varepsilon = \int_{-\infty}^{\varepsilon_V} 4\pi L^3 \left(\frac{2m_h^*}{h^2}\right)^{3/2} \sqrt{(\varepsilon_V - \varepsilon)} \exp\left(-\frac{\varepsilon_F - \varepsilon}{k_B T}\right) d\varepsilon \qquad (2.52)$$

Notice the integral limits in Equation (2.52) and compare it with the limits in Equation (2.29): in Equation (2.52), the energy is integrated from $-\infty$ to the top of the valence band (ε_V), whereas in Equation (2.29) the energy is integrated from the bottom of the conduction band (ε_C) to ∞.

The concentration of holes, which we denote by p, can be found by solving the integral of Equation (2.52) and dividing it by the volume. The result is:

$$p = \frac{N_h}{L^3} = n_d \exp\left(-\frac{\varepsilon_F - \varepsilon_V}{k_B T}\right) \qquad (2.53)$$

where n_d is given by:

$$n_d = \frac{2}{\left(\dfrac{h^2}{2\pi m_h^* k_B T}\right)^{3/2}} \qquad (2.54)$$

Comparing Equation (2.26) and Equation (2.54), we find that the only difference between n_c and n_d is that, while the former involves the effective mass of electrons, the latter involves the effective mass of holes.

An important property of semiconductors is that the product of n and p depends on the band gap energy ε_G and the temperature T. This property can be proved directly using Equation (2.47) and Equation (2.53):

$$n \cdot p = n_c n_d \exp\left(-\frac{\varepsilon_C - \varepsilon_F}{k_B T}\right) \exp\left(-\frac{\varepsilon_F - \varepsilon_V}{k_B T}\right) = n_c n_d \exp\left(-\frac{\varepsilon_C - \varepsilon_V}{k_B T}\right) \qquad (2.55)$$

Recall that, by definition, $\varepsilon_G = \varepsilon_C - \varepsilon_V$. Thus:

$$n \cdot p = n_c n_d \exp\left(-\frac{\varepsilon_G}{k_B T}\right) \qquad (2.56)$$

Keep in mind this important result.

So, we have an expression for n and an expression for p, and both of them (Equation (2.47) and Equation (2.53)) depend on the Fermi level ε_F. But where is the Fermi level in the band diagram? This question can be answered by noting that the concentrations of free electrons and holes must be the same. Recall that, at the temperature of $0\ K$, the valence band is fully occupied, and the conduction band is fully empty. Consequently, the concentrations of free electrons and holes are identically zero at the temperature of $0\ K$. As the temperature is increased, electrons in the valence band may be thermally excited to the conduction band, but each electron that was excited to the conduction band left behind a hole in the valence band. Therefore, the number of free electrons and the number of holes must be the same. Consequently, the concentrations are also the same, that is $n = p$. Thus:

$$n_c \exp\left(-\frac{\varepsilon_C - \varepsilon_F}{k_B T}\right) = n_d \exp\left(-\frac{\varepsilon_F - \varepsilon_V}{k_B T}\right) \tag{2.57}$$

Rearranging:

$$\exp\left(\frac{-2\varepsilon_F + \varepsilon_C + \varepsilon_V}{k_B T}\right) = \frac{n_c}{n_d} = \left(\frac{m_e^*}{m_h^*}\right)^{3/2} \tag{2.58}$$

Therefore:

$$-2\varepsilon_F + \varepsilon_C + \varepsilon_V = \frac{3}{2} k_B T \ln\left(\frac{m_e^*}{m_h^*}\right) \tag{2.59}$$

Consequently:

$$\varepsilon_F = \frac{\varepsilon_C + \varepsilon_V}{2} + \frac{3}{4} k_B T \ln\left(\frac{m_h^*}{m_e^*}\right) \tag{2.60}$$

Due to the typically weak interaction with the underlying atomic structure, the effective masses of electrons and holes are usually similar. Consequently, the second term in Equation (2.60) is small compared to the first term, which means that:

$$\varepsilon_F \approx \frac{\varepsilon_C + \varepsilon_V}{2} \tag{2.61}$$

We have thus reached the important conclusion that the Fermi level is in the middle of the band gap. One reason why semiconductors are so useful, however, is that it is easy to modify the Fermi level by a process called doping, which is the subject of the next section.

Box 7   Concentration of free electrons in intrinsic silicon

To get a feel for the order of magnitude of the free electrons concentration in intrinsic semiconductors, let us do a plug in numbers exercise for silicon.

We have found that the concentration of free electrons in an intrinsic semiconductor is given by:

$$n = n_c \exp\left(-\frac{\varepsilon_C - \varepsilon_F}{k_B T}\right)$$

Where

$$n_c = \frac{2}{\left(\frac{h^2}{2\pi m_e^* k_B T}\right)^{3/2}}$$

The effective mass of electrons in silicon is about $m_e^* = 1.09 \, m_0$, where m_0 is the electron mass. Let us make a list of the look up parameters and their units:

$$m_e^* = 1.09 \, m_0 = 1.09 \times 9.11 \times 10^{-31} \, [kg]$$

$$h = 6.626 \times 10^{-34} \left[\frac{m^2 kg}{s}\right]$$

$$k_B = 1.38 \times 10^{-23} \left[\frac{m^2 kg}{s^2 K}\right]$$

Before we plug in the numbers, let us check if n_c has units of concentration. Its units are:

$$\frac{1}{\left[\left(\frac{\left(\frac{m^2 kg}{s}\right)^2}{kg \frac{m^2 kg}{s^2 K} K}\right)^{3/2}\right]} = \frac{1}{\left[(m^2)^{3/2}\right]} = \frac{1}{[m^3]}$$

Thus, we have confirmed that n_c indeed has units of one over volume, which are the correct units for concentration.

Plugging in the numbers assuming room temperature ($T = 300 \, K$), we get:

$$n_c = \frac{2}{\left(\frac{h^2}{2\pi m_e^* k_B T}\right)^{3/2}} = \frac{2}{\left(\frac{\left(6.626 \times 10^{-34}\right)^2}{2\pi \times 1.09 \times 9.11 \times 10^{-31} \times 1.38 \times 10^{-23} \times 300}\right)^{3/2}}$$

Thus:

$$n_c = 2.85 \times 10^{25} \left[\frac{1}{m^3}\right]$$

(Continued)

(Continued)

Now we have to work out the exponential. To do that, it is more convenient to use units of electron volts, so I quote below Boltzmann constant in units of electron volts per kelvin:

$$k_B = 8.617 \times 10^{-5} \left[\frac{e.\,V.}{K} \right]$$

Silicon has a band gap of ~1.12 $e.\,V.$ at room temperature. If the Fermi level is close to the middle of the band gap, then $\varepsilon_C - \varepsilon_F \approx 1.12/2 = 0.56$ [$e.\,V.$]. Thus:

$$n = n_c \exp \left(-\frac{\varepsilon_C - \varepsilon_F}{k_B T} \right) \approx 2.815 \times 10^{25} \times \exp \left(-\frac{0.56}{8.617 \times 10^{-5} \times 300} \right)$$

Therefore:

$$n \approx 1.1 \times 10^{16} \left[\frac{1}{m^3} \right]$$

It is more common to quote the concentration in units of $\left[\frac{1}{cm^3} \right]$. Converting:

$$n \approx 11 \times 10^{9} \left[\frac{1}{cm^3} \right]$$

The experimental value is $\approx 9.65 \times 10^{9}$ $\left[\frac{1}{cm^3} \right]$, which is close to the value we found. The small difference comes mostly from the fact that the electron and hole effective masses in silicon are not exactly the same, so the Fermi level is not exactly at the middle of the band gap. Thus, a more accurate value for the distance $\varepsilon_C - \varepsilon_F$ is $\varepsilon_C - \varepsilon_F \approx 0.5637$. Now check that, if you plug in this value, you get $n \approx 9.67 \times 10^{9}$ $\left[\frac{1}{cm^3} \right]$.

2.6 Extrinsic semiconductors (doping)

Hitherto, we have been considering semiconductors in their "pure forms", which we denote by the term "intrinsic". One key advantage of semiconductors, however, is their amenability to control the concentrations of free electrons and holes by a process called "doping".

Doping is a process whereby atoms of impurity are inserted into the atomic structure of an otherwise "pure" semiconductor. Here, the term "impurity" is a bit misleading because it could convey the idea that anything alien to the semiconductor would do the job. This is far from true, however: atoms of impurity must be judiciously chosen according to the semiconductor that is being doped and according to the type of doping.

There are two types of doping: **n** doping and **p** doping. The former refers to the process whereby the concentration of free electrons is increased, whereas the latter refers to the process whereby the concentration of holes is increased.

A doped semiconductor is also called "extrinsic", as opposed to "intrinsic". Using a more direct nomenclature, we can designate extrinsic semiconductors by their doping type, as "**n** doped semiconductors" or "**p** doped semiconductors". There is a vast literature about the many doping techniques and descriptive videos showing the processes abound on the Internet. Given the easy accessibility to these materials, and their relative simplicity, we will not discuss these processes here. Instead, we will focus attention on the physics of doped semiconductors, thus paving the way to the next chapter, which focus attention on the physics of a device formed by bringing an **n** doped semiconductor and a **p** doped semiconductor together.

We begin the discussion by focusing attention on **n** doped semiconductors. An **n** doping impurity atom has one more valence electron than the semiconductor atom. For example, phosphorus has five valence electrons, whereas silicon has four valence electrons. Thus, phosphorus is an **n** doping atom of impurity for silicon. Since the chemical bonding between silicon and phosphorus involves only four electrons, there will be an extra electron from the phosphorus atom that does not participate in the bonding. It is this extra electron that contributes to an increase in the concentration of free electrons.

To understand how this extra electron ends up in the conduction band, consider the energy band of an **n** doped semiconductor, as shown in Figure 2.8. The insertion of an **n** doping impurity atom into the semiconductor creates an impurity orbital in the band gap. To conform with the literature, this impurity orbital is called an "impurity state" in Figure 2.8, but it is important to realize that these are not states of the total system; so, according to the nomenclature we adopted in Chapter 1, we need to call it an impurity orbital. Each impurity atom contributes with one impurity orbital, which is occupied by the extra electron that does not participate in the chemical bonding. Furthermore, the energy of the **n** doping impurity orbital lies close to the conduction band, as illustrated in Figure 2.8. Thus, electrons occupying these orbitals are easily thermally excited to the conduction band. Consequently, the **n** doping process populates the conduction band with extra free electrons, thus increasing the concentration of free electrons. Notice that, when an electron is thermally excited to the conduction band, it leaves behind a positively charged ionized impurity atom. Thus, these atoms are called "donors", because they "donate" one of their electrons to the conduction band.

What happens to the Fermi level when a semiconductor is **n** doped? As we have just seen, doping increases the concentration of free electrons. According to Equation (2.47), the concentration of free electrons depends on the "distance" between the Fermi level and the conduction band (through the term $\varepsilon_C - \varepsilon_F$). Still according to this equation, the smaller this separation

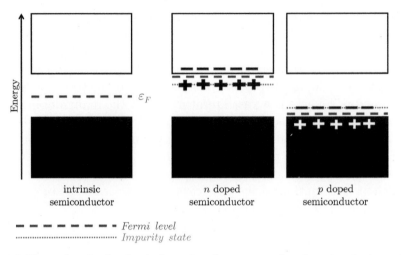

Energy

ε_F

intrinsic
semiconductor

n doped
semiconductor

p doped
semiconductor

– – – – – – – – · *Fermi level*
·················· *Impurity state*

Figure 2.8 Energy bands of an intrinsic semiconductor, an n doped semiconductor and a p doped semiconductor. Notice that n doping raises the Fermi level towards the conduction band, to which corresponds an increase in the concentration of free electrons; p doping, on the other hand, lowers the Fermi level towards the valence band, to which corresponds an increase in the concentration of holes.

is, the higher the concentration of free electrons is. Therefore, the Fermi level "moves up" towards the conduction band when the semiconductor is **n** doped, thus reducing the "distance" $\varepsilon_C - \varepsilon_F$.

This conclusion also makes sense from the point of view of the Fermi–Dirac distribution $f(\varepsilon)$. Recall that the probability of finding an orbital in the conduction band occupied is given by:

$$f(\varepsilon_{cb}) = \frac{1}{1 + \exp\left(\dfrac{\varepsilon_{cb} - \varepsilon_F}{k_B T}\right)}$$

Straightforward inspection of this equation shows that, the smaller is the "distance" between the energy of the orbital ε_{cb} and the Fermi level ε_F, then the higher is the probability that this orbital is occupied. Consequently, a higher ε_F (that is, a Fermi level closer to the conduction and) entails a higher probability $f(\varepsilon_{cb})$. Thus, we conclude that **n** doping increases the probability of occupancy of the orbitals in the conduction band.

Often in the literature one finds statements to the effect that an increase in the Fermi level *causes* an increase in the concentration of free electrons. It is important to notice that the increase of the Fermi level is not the *cause* of the increase of concentration of free electrons. Indeed, the physical cause of the increase of both Fermi level and concentration is the doping. And the *effect* of doping can be described either in terms of the Fermi level and/or the concentration of free electrons: both are equivalent descriptions of the same *effect*.

The physics of **p** doping is analogous to the physics of **n** doping. A **p** doping impurity atom is an atom that has one less valence electron than the atom of the semiconductor. For example, boron, which is a **p** doping atom for silicon, has only three valence electrons. As shown in Figure 2.8, the insertion of **p** doping impurity atoms creates impurity orbitals near the valence band. Electrons in the valence band are then easily thermally excited from the valence band to the impurity orbitals, thus leaving behind a hole in the valence band. Therefore, **p** doping increases the concentration of holes. According to Equation (2.53), the concentration of holes increases as the "distance" $\varepsilon_F - \varepsilon_V$ decreases. Therefore, the increase in the concentration of holes is associated with a lowering of the Fermi level towards the valence band. Finally, notice that an electron that is thermally excited to the impurity orbital negatively ionizes the impurity atom. Thus, these atoms are called "acceptors" because they accept an extra electron.

It is important to emphasize that doping does not alter the net charge of the semiconductor: both **n** and **p** doped semiconductors are still neutral materials. Indeed, as the impurity atoms are neutral (they have the name number of electrons and protons), so are the doped semiconductors. What happens is just an increase in the concentration of the free charge carriers, that is, an increase in the concentration of free electrons from **n** doping and of holes from **p** doping.

In the previous section we saw that an intrinsic semiconductor has the same concentration of free electrons and holes. This is no longer true for an extrinsic semiconductor. Indeed, an **n** doped semiconductor has a higher concentration of free electrons than holes, whereas a **p** doped semiconductor has a higher concentration of holes. Furthermore, typically, the differences in concentrations are large, so it can be affirmed that $n \gg p$ in an **n** doped semiconductor, whereas $p \gg n$ in a **p** doped semiconductor. Finally, notice that the product of n and p is not altered by doping, since this product does not depend on the Fermi level (Equation (2.55)). But if the product $n \cdot p$ is not altered by doping, and we have already seen that one of its terms increases with doping, then necessarily the other term must decrease. That means that the concentration of holes in an **n** doped semiconductor is lower than in the intrinsic semiconductor, whereas the concentration of free electrons in a **p** doped semiconductor is lower than in the intrinsic semiconductor. That makes sense: in **n** doping, the Fermi level goes up, which increases its distance to the valence band. But, according to Equation (2.53), a larger distance between the Fermi level and the valence band entails a lower concentration of holes. Likewise, in **p** doping, the Fermi level moves down, so its distance to the conduction band increases, and according to Equation (2.47) a larger distance between the Fermi level and the conduction band entails a lower concentration of free electrons (see Box 8).

Now we turn attention to the concentration of free charge carriers (free electrons and holes) and the Fermi level in doped semiconductors. These parameters obviously depend on the concentration of dopants. We denote the concentration of **n** doping atoms, or donor atoms, by N_d, where the

---

Box 8   Why does n doping affect the concentration of holes?

We have found that the concentration of free electrons increases in **n** doped semiconductors, and that this is due to the insertion of doping atoms, which contribute with extra electrons to the conduction band. Consequently, the Fermi level moves up towards the conduction band. Nothing very mysterious. But we also concluded something a bit less obvious: if the Fermi level goes up, then its separation from the valence band increases, and consequently the concentration of holes is reduced. But why? Why does **n** doping, which affects only the high energy part of the energy band, also affects the concentration of holes?

In the main text, we gave a "macroscopic" answer to this question, that is, we resorted to a macroscopic property of the system, as captured by the Fermi–Dirac distribution and the consequent equation for the concentration of holes. In other words, we blamed the Fermi level. Though correct, this explanation often leaves behind a nagging question: how exactly are the holes affected? Thus, it may be instructive to say a few words about the "microscopic" reason for the reduction in the concentration of holes.

A full description of the microscopic dynamic requires quantum mechanics, so it is outside the scope of this book. But we can still have a fairly good grasp of the logic behind it without delving into quantum mechanics. The logic goes like this: the probability that an electron is excited from the valence to the conduction band (thus leaving behind a hole) is given by the sum of the probabilities of transitions between the orbitals in the valence band, and all unoccupied orbitals in the conduction band (and we need quantum mechanics to work out these probabilities). When the semiconductor is **n** doped, there is an increase in the concentration of electrons in the conduction band, and, consequently, a reduction of unoccupied orbitals in the conduction band. Thus, **n** doping reduces the unoccupied orbitals, and consequently reduces the number of orbitals that contribute to the overall probability. So, it becomes less likely that an electron from the valence band is excited to the conduction band when the semiconductor is **n** doped just because there are less unoccupied orbitals in the conduction band for the electron to be excited into. Consequently, the concentration of holes is reduced when the semiconductor is **n** doped.

An analogous argument explains why **p** doping reduces the concentration of free electrons.

subscript d refers to "donor". Likewise, we denote the concentration of **p** doping atoms, or "accepting" atoms, by N_a.

The reasoning that led to the expressions for the concentration of free charge carriers (Equation (2.47) and Equation (2.53)) is still valid for doped semiconductors, so the equations are still valid. Recall, however, that the Fermi level in a doped semiconductor is no longer given by Equation (2.60) or Equation (2.61), since these equations were derived on the assumption that the concentrations of free electrons and holes are the same, which is no longer true in a doped semiconductor.

There is, however, a difference in point of view when using these equations (Equation (2.47) and Equation (2.53)) in an intrinsic and in a doped semiconductor. Recall that we deduced an expression for the Fermi level of an intrinsic semiconductor (Equation (2.60) or Equation (2.61)) that depends only on intrinsic parameters of the semiconductor (ε_C, ε_V and the effective masses). Thus, the Fermi level of an intrinsic semiconductor is also an intrinsic parameter, from which we can find the concentration of free charge carriers by substituting Equation (2.60) or Equation (2.61) into Equation (2.47) and Equation (2.53). In an intrinsic semiconductor, therefore, we infer the concentrations from the Fermi level. This point of view is somewhat flipped in a doped semiconductor, because in this case the concentrations of free charge carriers depend on the concentration of dopants, which is an external parameter controlled by the semiconductor scientist. For this reason, the more natural point of view in a doped semiconductor is to infer the Fermi level from the concentration of free charge carriers, and not the other way around.

The initial (intrinsic, before doping) concentration of charge carriers and the extra (extrinsic, coming from doping) concentration add together to form the total concentration. In most cases, however, the concentration due to doping is much higher than the intrinsic concentration. For this reason, it is a good approximation to treat the total concentration of free charge carriers as arising only from the doping process. Moreover, since each doping atom contributes with one free charge carrier, then the concentration of "extra" free charges carriers coincides with the concentration of dopants. Thus, in an **n** doped semiconductor, the concentration of free electrons n is almost identical to the concentration of dopants N_d ("almost identical" as opposed to "fully identical" due to the contribution of the intrinsic concentration), that is:

$$n \approx N_d$$

for **n** *doped semiconductors* (2.62)

Substituting Equation (2.62) into Equation (2.47), we get:

$$N_d \approx n_c \exp\left(-\frac{\varepsilon_C - \varepsilon_F}{k_B T}\right)$$

*for **n** doped semiconductors* (2.63)

The Fermi level can now be expressed in terms of the doping parameter N_d. Solving for ε_F in Equation (2.63):

$$\varepsilon_F = \varepsilon_C + k_B T \ln\left(\frac{N_d}{n_c}\right)$$

*for **n** doped semiconductors* (2.64)

Notice that, according to Equation (2.64), ε_F increases with N_d, as expected.

It is often convenient to express the concentration of free charges in a doped semiconductor in terms of the concentration of free charges in an intrinsic semiconductor. We denote the concentration of free electrons in an intrinsic semiconductor by n_i and the Fermi level of the intrinsic semiconductor by ε_{FI}. With the help of these definitions, Equation (2.47) can be recast as:

$$n = n_c \exp\left(-\frac{\varepsilon_C - \varepsilon_F}{k_B T}\right) = n_c \exp\left(-\frac{\varepsilon_C - \varepsilon_{FI} + \varepsilon_{FI} - \varepsilon_F}{k_B T}\right)$$

$$\therefore n = n_c \exp\left(-\frac{\varepsilon_C - \varepsilon_{FI}}{k_B T}\right) \exp\left(-\frac{\varepsilon_{FI} - \varepsilon_F}{k_B T}\right)$$ (2.65)

Noticing that, by definition, $n_i = n_c \exp\left(-\frac{\varepsilon_C - \varepsilon_{FI}}{k_B T}\right)$, Equation (2.65) reduces to:

$$n = n_i \exp\left(-\frac{\varepsilon_{FI} - \varepsilon_F}{k_B T}\right)$$ (2.66)

In the next section we will use Equation (2.66) instead of Equation (2.47). Notice that Equation (2.66) entails $n = n_i$ when $\varepsilon_F = \varepsilon_{FI}$, as expected.

We can also use Equation (2.66) to find an expression for the Fermi level in terms of n_i. Assuming once again that the doping is sufficiently high so that $n \approx N_d$, with the help of Equation (2.66) we have:

$$N_d \approx n_i \exp\left(-\frac{\varepsilon_{FI} - \varepsilon_F}{k_B T}\right)$$

*for **n** doped semiconductors* (2.67)

Solving for ε_F:

$$\varepsilon_F = \varepsilon_{FI} + k_B T \ln \left(\frac{N_d}{n_i} \right)$$

for n doped semiconductors (2.68)

Of course, Equation (2.68) and Equation (2.64) are equivalent. Recall, however, that these two equations are only valid for **n** doped semiconductors since they assume that $n \approx N_d$.

Notice, from Equation (2.68), that the larger the proportion of N_d with respect to n_i is, the larger the upwards shift of the Fermi level ε_F with respect to the intrinsic Fermi level ε_{FI} is (that is, the larger the shift with respect to the middle of the band gap is).

Box 9   Concentration of dopants and Fermi level

As an exercise, let us estimate the concentration of dopants necessary to move the Fermi level from the middle of the band gap up by a quarter of the band gap in silicon.

Assuming a band gap of 1.12 e. V, if the Fermi level is to move up by a quarter of the band gap, then we want $\varepsilon_F - \varepsilon_{FI} \approx 0.28$. Thus:

$$N_d \approx n_i \exp \left(-\frac{\varepsilon_{FI} - \varepsilon_F}{k_B T} \right) = n_i \exp \left(\frac{0.28}{k_B T} \right) = n_i \times 5.06 \times 10^4$$

That surely satisfies $N_d \gg n_i$.

In Box 7 we found $n_i \approx 1.1 \times 10^{10} \left[\frac{1}{cm^3} \right]$, so we need:

$$N_d \approx 5.5 \times 10^{14} \left[\frac{1}{cm^3} \right]$$

We can also find an expression for the concentration of holes in terms of the intrinsic concentration of holes p_i by rewriting Equation (2.53) as:

$$p = n_d \exp \left(-\frac{\varepsilon_F - \varepsilon_V}{k_B T} \right) = n_d \exp \left(-\frac{\varepsilon_F - \varepsilon_{FI} + \varepsilon_{FI} - \varepsilon_V}{k_B T} \right)$$

$$\therefore p = n_d \exp \left(-\frac{\varepsilon_{FI} - \varepsilon_V}{k_B T} \right) \exp \left(-\frac{\varepsilon_F - \varepsilon_{FI}}{k_B T} \right) \tag{2.69}$$

Therefore:

$$p = p_i \exp\left(-\frac{\varepsilon_F - \varepsilon_{FI}}{k_B T}\right) \tag{2.70}$$

Equation (2.70) and Equation (2.66) are completely general: they are valid for both intrinsic and extrinsic semiconductors, and for any type of doping. We have just seen, however, that in an **n** doped semiconductor we can make the approximation $n \approx N_d$, which leads to Equation (2.68) (or Equation (2.64)). Likewise, in a **p** doped semiconductor, the concentration of holes is approximately equal to the concentration of accepting impurity atoms:

$$p \approx N_a$$

*for **p** doped semiconductors* $\qquad\qquad$ (2.71)

Substituting Equation (2.71) into Equation (2.70), we find:

$$N_a = p_i \exp\left(-\frac{\varepsilon_F - \varepsilon_{FI}}{k_B T}\right)$$

*for **p** doped semiconductors* $\qquad\qquad$ (2.72)

The Fermi level in a **p** doped semiconductor can thus be expressed in terms of the concentration of dopants by solving for ε_F in Equation (2.72), resulting in:

$$\varepsilon_F = \varepsilon_{FI} - k_B T \ln\left(\frac{N_a}{p_i}\right)$$

*for **p** doped semiconductors* $\qquad\qquad$ (2.73)

Notice that, the larger the proportion of N_a with respect to p_i is, the larger the downwards shifting of the Fermi level with respect to the middle of the band gap is.

We have now acquired all the conceptual tools to finally engage with our target topic: the **p-n** junction, to which the next chapter is dedicated.

2.7   Suggestions for further reading

The content covered in this chapter is part of the discipline known as Solid State Physics. A standard world reference on this topic is Charles Kittel: *Introduction to Solid State Physics*.

2.8 Exercises

Exercise 1
Explain how the energy bands are formed from atomic orbitals.

Exercise 2
Explain how band occupation defines the electrical properties of a solid.

Exercise 3
We have seen that a hole is an absence of an electron. Thus, it can be said that, while an electron occupies an orbital, a hole "occupies an empty orbital". But there are lots of empty orbitals in the conduction band, so why can't we associate these empty orbitals with holes? Why only the empty orbitals of the valence bands are holes?

Exercise 4
We have seen that the equation giving the energy of the orbitals in the conduction band of a semiconductor (Equation (2.17)) is basically the same equation for the orbitals of a quantum well (Equation (1.58)), just with two modifications. However, whereas we cannot treat the orbitals of a quantum well as a continuum, the orbitals of the conduction band can be treated as forming a continuum (a band). Why is that? What parameter in these equations "decide" whether the energy differences between the orbital's energy levels are sufficiently small to justify treating them as a continuum?

Exercise 5

(a) Explain qualitatively the concept of density of states

(b) Derive an expression for the density of states of the system "electrons in a one-dimensional box".

Exercise 6
Derive the expression for the chemical potential of an ideal gas given in Exercise 10 of Chapter 1.

Exercise 7
Obtain the probability of finding an orbital at the bottom of the conduction band of silicon occupied at room temperature (assume $E_g = 1.12$ e. V.)

Exercise 8
Why is Equation (2.60) not valid for extrinsic semiconductors?

Exercise 9
Prove that the product of the concentration of free electrons and holes is not changed by doping.

If this product is not changed, what happens to the concentration of holes when a semiconductor is **n** doped? And what happens to the concentration of free electrons in a **p** doped semiconductor?

Exercise 10
Find an expression for the concentration of holes is an **n** doped semiconductor in terms of the intrinsic concentration and the concentration of doping atoms.

Exercise 11
Suppose that an intrinsic semiconductor is at temperature T. What is the mean energy of electrons in the conduction band of this semiconductor? Discuss the physical meaning of each term contributing to the mean energy, distinguishing between potential and kinetic energies.

Exercise 12
The probability of finding an electron at the bottom of the conduction band of a certain semiconductor is equal to the probability of finding an orbital unoccupied at the top of the valence band. Where is the Fermi level of this semiconductor?

3

Introduction to semiconductor devices: the p-n junction

Learning objectives

*In this chapter, we will collect everything learned in the previous chapters to derive the relationship between current and voltage in a **p-n** junction. But before embarking on the calculations in their full glory, we will pause to analyse the system qualitatively, to gain intuitions and to check what to expect from the calculations beforehand. One key feature of **p-n** junctions is their rectification property, which means that they are good conductors in one direction, and poor in the other.*

In this chapter, we apply the concepts learned in the previous chapters to study the physics of a **p-n** junction, which is the fundamental element of semiconductor devices. This is certainly the most important structure in electronics. With a good grasp of the physical and electrical properties of **p-n** junctions, you will already know how a diode works (a diode is just a **p-n** junction), how a solar cell works (another **p-n** junction), how a light emitting diode (LED) works (the very name of the thing already tells you that it is also a **p-n** junction) and you will be very well placed to understand how a transistor works (a transistor is just two **p-n** junctions).

We begin by studying the physical properties of a **p-n** junction in thermodynamic equilibrium. Next, we consider the properties of the junction when equilibrium is broken by the application of an external electrostatic potential difference to it. We conclude the chapter by deriving the relationship between current and voltage in a **p-n** junction, which is the famous Shockley equation.

3.1   p-n junction in thermodynamic equilibrium – qualitative description

As suggested by its name, a **p-n** junction is formed by the union of a **p** doped semiconductor and an **n** doped semiconductor. In practice, **p-n** junctions are formed by selective doping of a single substrate. However, for pedagogical reasons, we suppose that the junction is formed by bringing the two pieces of semiconductors together (I emphasize that this would never work in practice, but it is better to imagine it this way to understand the physics).

We studied a similar situation in Chapter 1, Figure 1.6d. In the system of Figure 1.6d, we noticed that the diffusion of charged elements creates a local charge imbalance that results in the creation of an internal electric field. This electric field (and its associated electrostatic potential difference) creates a drift current in the opposite direction of the diffusion current, and equilibrium is reached when the two currents cancel each other. In terms of the potentials, equilibrium is reached when the induced electrostatic potential difference compensates the initial chemical potential difference, so that the total potential, that is, the Fermi level, becomes constant across the materials.

What happens when a **p** doped semiconductor is brought into physical contact with an **n** doped semiconductor? Such a scenario is illustrated in Figure 3.1. As seen in Chapter 2, on the one hand, there is a high concentration of holes in the **p** doped semiconductor, which means that its Fermi level is low. On the other hand, there is a high concentration of free electrons in the **n** doped semiconductor – which means that its Fermi level is high. To differentiate between the Fermi levels of the two types, we denote the Fermi level of the **p** doped semiconductor as ε_{FP} and the Fermi level of the **n** doped semiconductor as ε_{FN}, as shown in Figure 3.1a. We learned in Chapter 1 that the Fermi level is the electrochemical potential. For now, however, there is no electrostatic potential difference across the semiconductors, so ε_{FP} and ε_{FN} are just the chemical potentials. We also learned in Chapter 1 that a difference in chemical potential results in a diffusion current. Since the chemical potential is higher in the **n** doped semiconductor then in the **p** doped semiconductor, when the two semiconductors are put in contact, electrons will diffuse from the **n** doped semiconductor to the **p** doped semiconductor. This makes sense: after all, the concentration of free electrons is higher in the **n** doped than in the **p** doped semiconductor.

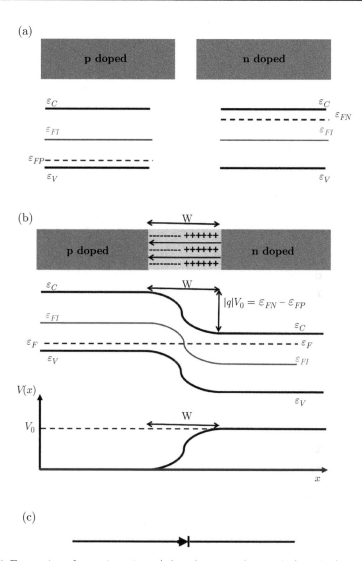

Figure 3.1 Formation of a p-n junction. a) doped semiconductors before the formation of the junction; ε_{FI} is the intrinsic Fermi level, ε_{FP} is the Fermi level of the p doped semiconductor and ε_{FN} is the Fermi level of the n doped semiconductor. b) p-n junction in thermodynamic equilibrium. Top figure: geometrical representation of the junction; middle figure: band diagram of the semiconductor junction in equilibrium – the energies are in units of joules or electron volts (not in units of volts); bottom figure: electrostatic potential in units of volts. Notice that, due to the negative electron charge, an increase in the electrostatic potential corresponds to a decrease in the energy; hence the downwards bending of the energy diagram. The region of width W, wherein the electric field is confined, is the "depletion region". Importantly, the Fermi level is constant across the band diagram, which is the signature of thermodynamic equilibrium. c) symbol for a p-n junction (a diode) used in circuits.

The situation of Figure 3.1 is similar to the situation of Figure 1.6d. In both cases, the two systems start out with a difference in chemical potential, which induces a diffusion of charge carriers; this diffusion causes an imbalance in the local charge, thus creating a kind of internal capacitor; the local charges create a local electric field, to which is associated a difference in electrostatic potential, which compensates the initial chemical potential difference; equilibrium is reached when the Fermi level (which is the electrochemical potential) becomes constant across the two materials. There is, however, an important difference between the systems: in Figure 1.6d there was only one charge carrier, but in the **p-n** junction there are two charge carriers: free electrons and holes. Therefore, there are two initial diffusion currents: electrons diffusing from the **n** doped semiconductor to the **p** doped semiconductor, and holes diffusing from the **p** doped semiconductor to the **n** doped semiconductor. This diffusion of holes is due to its higher concentration in the **p** doped semiconductor. Notice, however, the funny fact that the direction of diffusion of holes is from the material with lower chemical potential (the **p** doped semiconductor) to the material with higher chemical potential (the **n** doped semiconductor). Holes have this kind of inverse behaviour because they are, in reality, an absence of electrons in the valence band. So, keep in mind that **the direction of transport of holes is from the lower to the higher Fermi level**. Apart from this difference, the dynamics of transport of holes are essentially the same as the dynamics of electrons: the electric field induces a drift current of holes in the opposite direction of the diffusion, with the result that the net current of holes is zero in equilibrium.

At this stage, it may be useful to underline the difference between transport of elements and transport of charges. Recall that, if electrons diffuse from **n** to **p**, then the electric diffusion current is in the opposite direction, due to the negative charge of electrons (by electric diffusion current I mean the current of charges, in units of coulombs per unit of time). Thus, the direction of the diffusion electric current due to free electrons is from **p** to **n,** even though the electrons themselves diffuse from **n** to **p**. The diffusion of holes, at the same time, is from **p** to **n** and since the charge of holes is positive, its electric diffusion current is also from **p** to **n**. Thus, the diffusions of both electrons and holes cause a diffusion electric current (in coulombs per unit of time) whose direction is from **p** to **n**. A similar reasoning leads to the conclusion that the direction of the electric drift currents due to both charge carriers is from **n** to **p**.

The energy band diagram of a **p-n** junction in equilibrium is shown in Figure 3.1b. Notice the very important fact that the Fermi level ε_F is constant across the junction, which is the signature of thermodynamic equilibrium, as has been emphasized since Chapter 1. Recall that neither the chemical potential on its own, nor the electrostatic potential on its own, are constant across the junction: but the total potential, that is, the Fermi level, is indeed constant across the junction.

Another prominent feature of the energy band is that it bends in the middle. Such bending is necessary to align ε_{FP} and ε_{FN}, thus resulting in the constant ε_F. In other words, the electric field creates an electrostatic potential difference V_0 in units of volts (see Figure 3.1b), to which is associated an electrostatic potential difference $-|q|V_0$ in units of joules (the negative sign is due to the electron charge) in the band diagram. This means that the electric field lowers the energy on the **n** side by $|q|V_0$ with respect to the **p** side. As we already know, equilibrium is reached when the induced electrostatic potential $|q|V_0$ compensates the initial chemical potential difference $\varepsilon_{FN} - \varepsilon_{FP}$, that is, when $|q|V_0 = \varepsilon_{FN} - \varepsilon_{FP}$, as shown in Figure 3.1b.

Often, students having a first contact with the theory find this energy bending somewhat puzzling. There is, however, nothing very mysterious about it. The mundane nature of the bending can be illustrated by an analogous system, as shown by the artistic masterpiece exhibited in Figure 3.2. Consider a sketch of the internal energy of a person lying down, as shown on the left-hand side of Figure 3.2. The internal energy of the person is the energy of all atoms and molecules constituting this person, and it is somewhat analogous to the band diagram of free electrons before the junction is formed: as there is no gravitational potential gradient across the person, likewise there is no electrostatic potential gradient in the band diagram before the junction is formed (see Figure 3.1a). Next, suppose that the person leans on a slope, as shown on the right-hand side of Figure 3.2. Now, there is

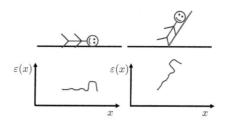

Figure 3.2 Gravitational analogy of the effect of the electrostatic potential on the energy band. Left side: representation of the internal energy of a person lying down, a situation where there is no gravitational potential difference across the person; this situation is analogous to the energy bands before the formation of the junction, where there is no electrostatic potential difference across the band. Right side: energy of the same person, but now in a situation where there is a gravitational potential gradient across the person bending the internal energy; this situation is analogous to the band diagram after the formation of the junction, where there is an electrostatic potential difference bending the energy band.

an extra gravitational contribution to the internal energy of the person. Furthermore, this contribution is not uniform: the head is at a higher gravitational potential energy than the feet. Consequently, gravity "bends" the internal energy profile. An analogous situation is found in the band diagram: the electric field induces an electrostatic potential gradient, thus bending the energy diagram (as in Figure 3.1b).

Still referring to Figure 3.1b, notice the region where the local charges accumulate, forming the "internal capacitor" at the interface between the **p** and **n** sides. This region, to which we attribute the width **W**, is called the "depletion region". Such terminology is due to the lack of free charge carriers in this region, that is, the region is "depleted" of free charges. Indeed, the free charges diffused to the other side, leaving behind the ionized doping atoms with a charge imbalance. These ionized atoms, therefore, constitute the local charges that form the "capacitor" (recall that the doping atoms are initially neutral, but they become ionized due to thermal excitation to the conduction band for **n** doping and thermal excitation from the valence band for **p** doping). Obviously, these are not free charges: the doping atoms are bound to the semiconductor. In the literature, it is often said, using a somewhat flippant terminology, that these ionized atoms were left "naked" by the free charges that diffused to the other side, thus exposing the bound charge of the ions. Never forget, however, that the semiconductor is still neutral: for each negative local charge on one side, there is a positive one on the other side.

3.2 p-n junction in thermodynamic equilibrium – quantitative description

The main qualitative features of **p-n** junctions in equilibrium were covered in the previous section. Now we proceed to a more rigorous treatment. As a comprehensive description of all electrical properties of **p-n** junctions is likely to overwhelm students at this stage, we focus attention on three of its most important properties: the built-in potential V_0, the electric field and the width of the depletion region. Our agenda for this section is: we start by proving the relation $|q|V_0 = \varepsilon_{FN} - \varepsilon_{FP}$, then we derive an expression for the electric field in terms of the concentration of doping atoms and an expression for the width W of the depletion region. I am confident that the student who grasps the logic of these derivations is well placed to understand other properties that were left out, as the internal capacitance, for example.

In Chapter 1 we concluded that the electrostatic potential difference must compensate the initial chemical potential difference for the system to reach equilibrium. This requirement, when applied to the **p-n** junction, entails that $|q|V_0 = \varepsilon_{FN} - \varepsilon_{FP}$. It is instructive to prove this relation more formally, as it involves connecting the equilibrium condition for currents with the equilibrium condition for potentials. Thus, we pursue the proof of the equality $|q|V_0 = \varepsilon_{FN} - \varepsilon_{FP}$ by invoking the condition of zero net current, that is, the contributions of drift and diffusion currents must cancel each other in equilibrium. This requirement applies for both free electrons and holes independently, and we can choose either for the calculations. I choose to do them using the current of holes, due to its positive charge.

To begin, let us recall the mathematical expressions for the diffusion and drift currents, which we derived in section 1.10. Denoting the drift current by J_{drift}, we have (Equation (1.77))

$$J_{drift} = \rho \cdot \langle v \rangle = |q| \cdot p \cdot \langle v \rangle \tag{3.1}$$

Notice that, since we are dealing with an electric current, then ρ must be the concentration of charges (in units of coulombs per unit of volume). Since p is the concentration of holes (that is, the number of holes per unit of volume), then it follows that $\rho = |q| \cdot p$.

The diffusion current, which we denote by J_{dif}, is given by (Equation (1.90)):

$$J_{dif} = -D \cdot \frac{\partial \rho}{\partial x} = -|q| \cdot D \cdot \frac{\partial p}{\partial x} \tag{3.2}$$

Notice that, while in section 1.10 we adopted a horizontal z-axis, now we are adopting a horizontal x-axis. Obviously, such a difference has no physical content, and only reflects the most common choices in the literature.

Before applying the equilibrium condition, we need to tweak Equation (3.1) and Equation (3.2) a little bit. Starting with Equation (3.1), according to Ohm's law, the mean velocity $\langle v \rangle$ that appears in the drift current is proportional to the electric field. The proportionality constant is the "mobility" of the charge carriers, which is usually denoted by the symbol μ in the literature on semiconductors (this is the same symbol we used for the electrochemical potential in Chapter 1, so beware not to mix them up). Writing $\langle v \rangle$ in terms of the electric field $E(x)$, Equation (3.1) reads:

$$J_{drift} = |q| \cdot p(x) \cdot \langle v \rangle = |q| \cdot p(x) \cdot \mu \cdot E(x) \tag{3.3}$$

As the electric field and concentration of holes are not uniform, it is convenient to express them explicitly as a function of the position x, as we have done in Equation (3.3).

Among his other conspicuous accomplishments, the great Albert Einstein once proved that the diffusion coefficient D and the mobility μ are not independent of each other. As you will prove yourself in the exercise list at the end of this chapter, they are related as:

$$|q|D = \mu \cdot k_B \cdot T \tag{3.4}$$

With the help of Equation (3.4), the diffusion current can be recast as:

$$J_{dif} = -|q| \cdot D \cdot \frac{\partial p}{\partial x} = -\mu \cdot k_B \cdot T \cdot \frac{\partial p}{\partial x} \tag{3.5}$$

Now we apply the equilibrium condition to the total current:

$$J_{drift} + J_{dif} = 0 \tag{3.6}$$

From which it follows that:

$$|q| \cdot p(x) \cdot \mu \cdot E(x) - \mu \cdot k_B \cdot T \cdot \frac{\partial p}{\partial x} = 0 \tag{3.7}$$

Solving for the electric field:

$$E(x) = \frac{k_B T}{|q|} \cdot \frac{1}{p(x)} \cdot \frac{\partial p}{\partial x} \tag{3.8}$$

Let us pause a minute to take stock of the situation. Our goal is to prove that $|q|V_0 = \varepsilon_{FN} - \varepsilon_{FP}$. In the previous chapter, we saw that the concentrations of free charge carriers depend on the Fermi levels (Equation (2.66) and Equation (2.70)). Thus, according to Equation (2.70), the hole concentration $p(x)$ is a function of the Fermi level. Therefore, the right-hand side of Equation (3.8) depends on the Fermi level through $p(x)$. The voltage V_0, on the other hand, can be found by integrating the electric field across the depletion region. To perform this integration, we need a coordinate axis to designate the depletion region. As shown in Figure 3.3, we denote the edges of the depletion region as the points $-x_p$ and x_n.

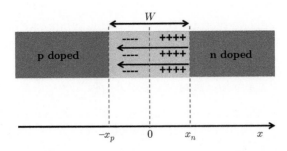

Figure 3.3 Representation of the coordinate axis used to integrate the electric field. The point $-x_p$ designates the beginning of the depletion region, whereas the point x_n designates the end of the depletion region.

As discussed earlier, the electric field is confined inside the depletion region (for the same reason that the electric field of an ideal parallel plate capacitor is confined between the plates). The total potential difference due to the electric field can then be found by integration over the depletion region. Thus:

$$V(x_n) - V(-x_p) = -\int_{-x_p}^{x_n} E(x)\,dx \tag{3.9}$$

By definition, the built-in potential V_0 is the potential difference due to the internal electric field, that is $V_0 = V(x_n) - V(-x_p)$ – see Figure 3.1b. Thus:

$$V_0 = -\int_{-x_p}^{x_n} E(x)\,dx \tag{3.10}$$

Substituting Equation (3.8) into Equation (3.10), we find:

$$V_0 = -\int_{-x_p}^{x_n} \frac{k_B T}{|q|} \cdot \frac{1}{p(x)} \cdot \frac{\partial p}{\partial x}\,dx$$

$$\therefore V_0 = -\frac{k_B T}{|q|}\left[\ln p(x_n) - \ln p(-x_p)\right] \tag{3.11}$$

We have thus reached the important conclusion that:

$$V_0 = \frac{k_B T}{|q|}\ln\left[\frac{p(-x_p)}{p(x_n)}\right] \tag{3.12}$$

Remember that we could have followed a similar path using the currents of free electrons. If we had chosen to do it with free electrons, we would have concluded that:

$$V_0 = \frac{k_B T}{|q|}\ln\left[\frac{n(x_n)}{n(-x_p)}\right] \tag{3.13}$$

So, in our path to prove that $|q|V_0 = \varepsilon_{FN} - \varepsilon_{FP}$, we encountered two important relations for the potential difference, given by Equation (3.12) and Equation (3.13). Combining these two equations, we conclude that

$$n(-x_p) \cdot p(-x_p) = n(x_n) \cdot p(x_n) \tag{3.14}$$

According to Equation (3.14) the products $n \cdot p$ at the edges of the depletion region are the same. Recall that the concentrations depend on the difference between the Fermi level and the intrinsic Fermi level (Equation (2.66) and Equation (2.70)). The Fermi level is constant across the entire semiconductor (since it is in equilibrium), and the intrinsic Fermi level is constant outside the depletion region (since there is no bending – see Figure 3.1b). Thus, Equation (3.14) entails that the product $n \cdot p$ is the same on both sides of the junction. In the previous chapter, we noticed that the product $n \cdot p$ does not depend on doping, so it makes sense that they are the same on both sides of the junction.

With the help of Equation (2.70), Equation (3.12) can be expressed as:

$$V_0 = \frac{k_B T}{|q|} \ln\left[\frac{p(-x_p)}{p(x_n)}\right] = \frac{k_B T}{|q|} \ln\left[\frac{p_i \exp\left[-\frac{\varepsilon_F - \varepsilon_{FI}(-x_p)}{k_B T}\right]}{p_i \exp\left[-\frac{\varepsilon_F - \varepsilon_{FI}(x_n)}{k_B T}\right]}\right]$$

$$= \frac{k_B T}{|q|} \ln\left[\frac{\exp\left[-\frac{\varepsilon_F - \varepsilon_{FI}(-x_p)}{k_B T}\right]}{\exp\left[-\frac{\varepsilon_F - \varepsilon_{FI}(x_n)}{k_B T}\right]}\right] \tag{3.15}$$

Notice that the intrinsic Fermi level $\varepsilon_{FI}(x)$ is a function of position x, since it varies across the junction (see Figure 3.1b). Accordingly, $\varepsilon_{FI}(-x_p)$ is the intrinsic Fermi level at the beginning of the depletion region, and $\varepsilon_{FI}(x_n)$ is the intrinsic Fermi level at the end of the depletion region – see Figure 3.1b and Figure 3.3. Furthermore, since the intrinsic Fermi level does not vary outside of the depletion region, then its value across the entire **p** region is also $\varepsilon_{FI}(-x_p)$; likewise, the value across the **n** region is also $\varepsilon_{FI}(x_n)$. Therefore, according to Equation (3.15), V_0 depends on the difference between the Fermi level ε_F and the intrinsic Fermi level ε_{FI} at the two sides of the **p-n** junction (outside the depletion region). Outside the depletion region, however, this difference is the same as in the doped semiconductors before the formation of the junction (compare Figure 3.1a and Figure 3.1b). In mathematical terms, this means that, in the region of the **p** side ($x \leq -x_p$), we have:

$$\left(\varepsilon_F - \varepsilon_{FI}\right)_{JUNCTION\ FOR\ x \leq -x_p} = \left(\varepsilon_{FP} - \varepsilon_{FI}\right)_{P\ DOPED\ BEFORE\ JUNCTION} \tag{3.16}$$

The left-hand side of Equation (3.16) is the difference $\varepsilon_F - \varepsilon_{FI}$ in the junction for the region $x \leq -x_p$. The right-hand side of Equation (3.16) is the difference $(\varepsilon_{FP} - \varepsilon_{FI})$ of a **p** doped semiconductor, as shown in Figure 3.1a.

Similarly, for the **n** side ($x \geq x_n$), we have:

$$\left(\varepsilon_F - \varepsilon_{FI}\right)_{JUNCTION\ FOR\ x \geq x_n} = \left(\varepsilon_{FN} - \varepsilon_{FI}\right)_{N\ DOPED\ BEFORE\ JUNCTION} \tag{3.17}$$

With the help of Equation (3.16) and Equation (3.17), Equation (3.15) can be written as:

$$V_0 = \frac{k_B T}{|q|} \ln \left[\frac{\exp\left[-\frac{\varepsilon_F - \varepsilon_{FI}(-x_p)}{k_B T} \right]_{JUNCTION}}{\exp\left[-\frac{\varepsilon_F - \varepsilon_{FI}(x_n)}{k_B T} \right]_{JUNCTION}} \right]$$

$$= \frac{k_B T}{|q|} \ln \left[\frac{\exp\left[-\frac{\varepsilon_{FP} - \varepsilon_{FI}}{k_B T} \right]_{P\ DOPED\ BEFORE\ JUNCTION}}{\exp\left[-\frac{\varepsilon_{FN} - \varepsilon_{FI}}{k_B T} \right]_{N\ DOPED\ BEFORE\ JUNCTION}} \right] \qquad (3.18)$$

But the intrinsic Fermi level was the same in the **p** and **n** doped semiconductors before the junction was formed (see Figure 3.1a). Thus, Equation (3.18) reduces to:

$$V_0 = \frac{k_B T}{|q|} \ln \left[\exp\left[-\frac{\varepsilon_{FP} - \varepsilon_{FN}}{k_B T} \right] \right] \qquad (3.19)$$

From which we conclude that:

$$V_0 = \frac{\varepsilon_{FN} - \varepsilon_{FP}}{|q|} \qquad (3.20)$$

As expected. We have thus completed our first task.

Importantly, notice that the built-in electrostatic potential difference V_0 can be found in terms of doping parameters through Equation (3.20): the information about doping enters Equation (3.20) through the terms ε_{FN} and ε_{FP}, which depend on N_d and N_a through Equation (2.68) and Equation (2.73), respectively.

Now we move on to the second task our agenda: to find the electric field in terms of N_d and N_a. As discussed earlier, the electric field is caused mainly by the naked impurity ions. To be sure, there is also the contribution of the diffused charges: free electrons that diffuse from **n** to **p** suffer the process of recombination in the **p** region, that is, they fall back from the conduction band to the valence band, thus "annihilating" a free electron and a hole. This process also causes a charge imbalance, but we can dismiss its contribution to the electric field because the recombined charges are not concentrated in a small and well-defined region, as is the case for the impurity ions. Recombination also happens for holes that diffused from **p** to **n**, and we can also dismiss its contribution to the electric field for the same reason. So, it makes sense to look for an expression for the field in terms of only the concentrations of naked impurity ions.

A semiconductor may be doped in different ways, using various doping profiles. For simplicity, we assume that the doping profile is uniform, but the same logic applies to any kind of doping (though the math may get more cumbersome for exotic doping profiles). For uniform doping, the concentrations N_a of **p** dopants is constant across the **p** side of the junction, and the concentration N_d of **n** dopants is constant across the **n** side. At the interface, there is an abrupt change from N_a to N_d. Of course, only the naked doping ions contribute to the electric field (the others are covered by the free charges, so there is no local charge accumulation). As shown in Figure 3.4, we place the interface of dopants at the origin of the coordinate system. Notice that each type of charge is concentrated between the edge of the depletion region and the origin of the coordinate system.

The electric field caused by the ionic charge density can be found from the laws of electrostatics; more specifically, from Gauss's law, according to which:

$$\nabla \cdot \vec{E} = \frac{\rho}{\epsilon} \qquad (3.21)$$

where ϵ is the dielectric constant of the semiconductor and ρ is the charge density.

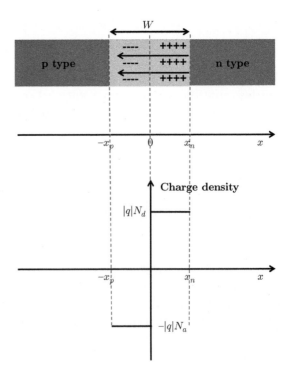

Figure 3.4 Spatial distribution of charge density.

Since our problem is unidimensional, the divergence in Gauss's law is reduced to a single partial derivative. Thus:

$$\frac{\partial E}{\partial x} = \frac{\rho}{\epsilon} \tag{3.22}$$

For uniform doping, the charge density is given by (see Figure 3.4):

$$\rho = -|q|N_a \quad for \quad -x_p \le x < 0$$

and

$$\rho = |q|N_d \quad for \quad 0 < x \le x_n \tag{3.23}$$

Therefore:

$$\frac{\partial E}{\partial x} = \frac{-|q|N_a}{\epsilon} \quad for \quad -x_p \le x < 0$$

and

$$\frac{\partial E}{\partial x} = \frac{|q|N_d}{\epsilon} \quad for \quad 0 < x \le x_n \tag{3.24}$$

Integrating Equation (3.24), we find:

$$E(x) = E(0) - \frac{|q|N_a}{\epsilon}x \quad for \quad -x_p \le x < 0$$

and

$$E(x) = E(0) + \frac{|q|N_d}{\epsilon}x \quad for \quad 0 < x \le x_n \tag{3.25}$$

As discussed earlier, the field outside the depletion region is null. Therefore, the boundary conditions for Equation (3.25) are $E(-x_p) = E(x_n) = 0$. Equation (3.25), with these boundary conditions, implies a triangular electric field profile, as shown in Figure 3.5. Notice that the field amplitude is negative, which is in agreement with Figure 3.4: the field points to the left of the x-coordinate axis (because the positive charges are on the right-hand side, and the negative charges are on the left-hand side).

The boundary conditions $E(-x_p) = E(x_n) = 0$ entail that:

$$|E(0)| = \frac{|q|N_a}{\epsilon}x_p = \frac{|q|N_d}{\epsilon}x_n \tag{3.26}$$

Equation (3.25) and Equation (3.26) give the electric field profile in terms of the concentration of doping atoms (N_a or N_d), but they still depend on the width of the depletion region through the parameters x_p and x_n. Thus, to conclude our second task, we still need to express these parameters in terms of N_a and N_d. To do so, first notice that Equation (3.26) entails that $N_a x_p = N_d x_n$.

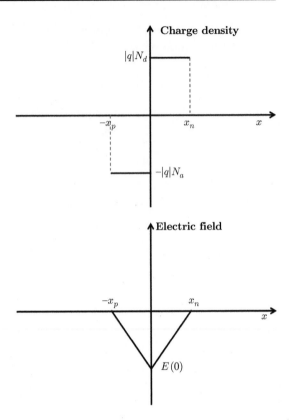

Figure 3.5 Spatial dependence of charge density and electric field.

The width W of the depletion region, on the other hand, is $W = x_n + x_p$ (see Figure 3.4 and Figure 3.5). Therefore:

$$W = x_n + x_p = x_n \left(1 + \frac{N_d}{N_a}\right) \tag{3.27}$$

Notice that, since:

$$V_0 = -\int_{-x_p}^{x_n} E(x)\,dx$$

Then V_0 is the area of the triangle in Figure 3.5, that is:

$$V_0 = \frac{(x_n + x_p)\,|\,E(0)\,|}{2} = \frac{W\,|\,E(0)\,|}{2}$$

Thus, with the help of Equation (3.26), we find:

$$V_0 = \frac{W}{2}|E(0)| = \frac{W}{2}\frac{|q|N_a}{\epsilon}x_p \ \ or \ \ V_0 = \frac{W}{2}|E(0)| = \frac{W}{2}\frac{|q|N_d}{\epsilon}x_n \tag{3.28}$$

Combining Equation (3.27) and Equation (3.28), we find:

$$W = \sqrt{\frac{2 \cdot \epsilon \cdot V_0}{|q|}\left(\frac{N_a + N_d}{N_a \cdot N_d}\right)}\tag{3.29}$$

The parameters x_p and x_n can be found in terms of the doping parameters by substituting Equation (3.29) into Equation (3.28) and solving for x_p or x_n. This completes our second task. Furthermore, our third task, which was to find an expression for the width of the depletion region, has also been completed in Equation (3.29) (recall that we have already found an expression for V_0 in terms of the doping parameters through Equation (3.20), assisted by Equation (2.68) and Equation (2.73).

The main features of the **p-n** junction in thermodynamic equilibrium have now been established: all three parameters were expressed in terms of the concentrations of doping atoms N_a and N_d. These expressions allow the engineer to choose the concentrations N_a and N_d according to the desired electrical properties.

The concepts developed in this section pave the way to our main objective, which is to understand how the **p-n** junction behaves when an external voltage is applied to it, as in a circuit. Thus, the condition of interest is no longer one of thermodynamic equilibrium, since this equilibrium is broken by the application of an external electrostatic potential difference. To understand its behaviour out of the equilibrium condition, however, we need to find out what happens to the Fermi level when the system is no longer at thermodynamic equilibrium. This is the subject of the next section.

3.3   Systems out of thermodynamic equilibrium: the *quasi*-Fermi levels

In Chapter 2, we learned that the concentrations of free electrons n and holes p are given, respectively, by:

$$n = n_c \exp\left(-\frac{\epsilon_C - \epsilon_F}{k_B T}\right) \quad or \quad n = n_i \exp\left(-\frac{\epsilon_{FI} - \epsilon_F}{k_B T}\right)\tag{3.30}$$

and

$$p = n_d \exp\left(-\frac{\epsilon_F - \epsilon_V}{k_B T}\right) \quad or \quad p = p_i \exp\left(-\frac{\epsilon_F - \epsilon_{FI}}{k_B T}\right)\tag{3.31}$$

In Chapter 2, we also showed that the product $n \cdot p$ does not depend on the Fermi level. Consequently, it is not altered by doping. Indeed, as can be straightforwardly proved from the two equations above, $n \cdot p = n_i \cdot p_i$. Thus, if n is made higher than n_i through **n** doping, then necessarily **n** doping also makes p lower than p_i. Likewise, if p is made higher than p_i through **p** doping, then necessarily **p** doping also makes n lower than n_i. In short: doping increases the concentration of one type of carrier while lowering the concentration of the other type.

Now, consider an intrinsic semiconductor (so $n = n_i$ and $p = p_i$). In Chapter 2 we learned that n_i and p_i depend on the band gap and the temperature. Therefore, if the temperature is fixed, then n_i and p_i are also fixed (and recall that they are also the same: $n_i = p_i$). Now suppose you shine light onto an intrinsic semiconductor. If the energy of the photons is higher than the band gap (the energy of each photon is proportional to the frequency of the radiation), then an electron in the valence band can be excited to the conduction band upon absorption of a photon. Thus, each one of these electrons become a free electron in the conduction band, and each one of these electrons leave a hole in the valence band. Therefore, shining light onto the semiconductor increases the concentration of free electrons and holes **at the same time!!**

What are we to do with this strange situation? We saw in Chapter 2 that an increase in the concentration of free electrons corresponds to an increase in the Fermi level. But we also saw that an increase in the concentration of holes corresponds to a decrease in the Fermi level. What happens to the Fermi level when these two concentrations are increased simultaneously?

This situation is a paradigmatic example of a semiconductor that is no longer at thermodynamic equilibrium. Indeed, an extra energy has been injected into the semiconductor by shinning light onto it, thus pushing it away from the equilibrium condition. It is precisely this extra energy that accounts for the simultaneous increase in the concentrations of free electrons and holes. Since the increase in the concentration of free electrons corresponds to an increase in the Fermi level and an increase in the concentration of holes corresponds to a decrease in the Fermi level, there results a split in the Fermi level, as illustrated in Figure 3.6. Thus, out of equilibrium, there is one Fermi level for free electrons, and another Fermi level for holes. But since this happens at a condition that is no longer of thermodynamic equilibrium, we use the term *quasi*-Fermi level. **The splitting of the Fermi level into *quasi*-Fermi levels is the signature of a system out of thermodynamic equilibrium.** Notice that, contrary to the doping process, the product $n \cdot p$ is changed when the system is pushed away from equilibrium. Indeed, if both n and p increases simultaneously, so does their product.

The *quasi*-Fermi levels play a similar role to the Fermi level and they can be considered as an extension of this concept to include situations out of equilibrium. The term *quasi* denotes a condition that is not equilibrium, but that it varies slowly enough so the system can be considered as in *quasi*-equilibrium. For example, if the illumination does not vary too fast in time, then the illuminated semiconductor can be considered to be in *quasi*-equilibrium.

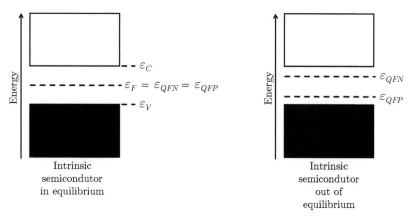

Figure 3.6 Illustration of the Fermi level splitting into *quasi*-Fermi level.

In a condition of *quasi*-equilibrium, Equation (3.30) and Equation (3.31) are still valid, but we need to substitute the *quasi*-Fermi levels for the Fermi level. Denoting the *quasi*-Fermi level for free electrons as ε_{QFN} and the *quasi*-Fermi level for holes as ε_{QFP}, these equations read:

$$n = n_c \exp\left(-\frac{\varepsilon_C - \varepsilon_{QFN}}{k_B T}\right) \quad or \quad n = n_i \exp\left(-\frac{\varepsilon_{FI} - \varepsilon_{QFN}}{k_B T}\right) \tag{3.32}$$

and

$$p = n_d \exp\left(-\frac{\varepsilon_{QFP} - \varepsilon_V}{k_B T}\right) \quad or \quad p = p_i \exp\left(-\frac{\varepsilon_{QFP} - \varepsilon_{FI}}{k_B T}\right) \tag{3.33}$$

Illumination is not the only way to break thermodynamic equilibrium in a semiconductor. Indeed, the most common way is to apply an external voltage to it. But the principle is the same: it does not matter how equilibrium is broken: out of equilibrium there will be a splitting of the Fermi level into *quasi*-Fermi levels. In the next sections, we apply these concepts to study the behaviour of the **p-n** junction when equilibrium is broken by an external voltage.

3.4 Qualitative description of the relationship between current and voltage in a p-n junction

In the next section, we will derive an expression for the relationship between current and voltage in a **p-n** junction. This is the famous Shockley equation, and it is the crowning result of this book. Before we begin the derivation of Shockley equation, however, we need to understand qualitatively what happens to a **p-n** junction when an external voltage is applied to it. This is the objective of this section.

An external voltage can be applied to a **p-n** junction in two ways. On the one hand, when the voltage on the **p** side is higher than on the **n** side, the junction is said to be forward biased (see Figure 3.7a). On the other hand,

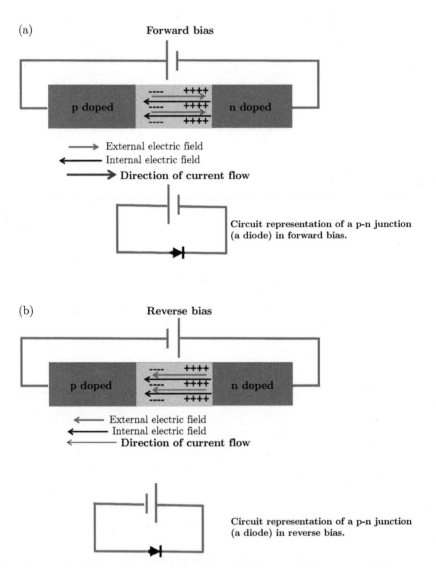

Figure 3.7 Types of polarization bias in a p-n junction. a) Forward bias: the voltage on the p side is higher than the voltage on the n side; the p-n junction is a good conductor in forward bias. b) Reverse bias: the voltage on the p side is lower than on the n side; the p-n junction is a poor conductor in reverse bias.

when the voltage on the **p** side is lower than on the **n** side, the junction is said to be reverse biased (see Figure 3.7b).

To understand what happens when the junction is biased, recall the condition of thermodynamic equilibrium: without any bias, the diffusion current arising from the chemical potential gradient (that is, the difference in the concentrations of charge carriers between the two sides) is cancelled by the drift current caused by the built-in electrostatic potential difference (which is due to the internal electric field). In terms of potentials, we have learned that the electrostatic potential, which is responsible for the drift current, cancels the difference in the chemical potential, which is responsible for the diffusion current. Thus, one can say that the electrostatic potential acts as a "barrier" to the diffusion of charges – the charges that try to diffuse to the other side are pulled back by the electric field. But, if an external voltage is applied, then there will be an external electric field, which will either oppose the internal electric field (forward bias) or align with the internal electric field (reverse bias). Consequently, the electrostatic potential "barrier" is either lowered or increased by the application of an external voltage.

If the junction is forward biased, as shown in Figure 3.7a, then the external electric field opposes the internal electric field. Consequently, the total field in the depletion region is reduced, which means that the electrostatic potential barrier and the drift current are both lowered. As the drift current no longer fully counterbalances the diffusion current, a net current flows in the same direction as the diffusion current, that is, from **p** to **n**.

Should we expect this net current to be high or low? On the one hand, we already know that there is a steep concentration gradient in the **p-n** junction: there is a high concentration of free electrons in the **n** side, and a low concentration of free electrons in the **p** side; likewise, there is a high concentration of holes in the **p** side, and a low concentration of holes in the **n** side. On the other hand, we also know that the diffusion current is proportional to the concentration gradient (Equation (1.90) or Equation (3.2)). Therefore, we expect the net current to be high due to the steep concentration gradient. Thus, we conclude that the **p-n** junction is a good conductor when it is forward biased.

In reverse bias, however, the external electric field is parallel to the internal electric field, thus resulting in a higher total field, as shown in Figure 3.7b. Consequently, the electrostatic potential barrier increases when the junction is reverse biased. Furthermore, the increase in the electric field (and potential barrier) increases the drift current, thus allowing a net current to flow in the same direction of the drift current, that is, from the **n** to the **p** side.

Should we expect this net current to be high or low? Recall that, in equilibrium, the drift current is essentially the current arising from the electric field pulling back the charges that were tyring to diffuse to the other side. But if the drift current is increased, then there must be others, "extra", charges contributing to the current. Where are these "extra" charges coming from? Well, they can only come from the **p** and **n** sides. But recall that the direction of the drift current is from the **n** side to the **p** side, which means that this current is constituted by holes being pulled from the **n** side to the **p** side, and electrons being pulled from the **p** side to the **n** side. Thus, these "extra" charges required to increase the drift current beyond its equilibrium condition are the holes in the **n** side and the free electrons in the **p** side. But the concentrations of holes in the **n** side and free electrons in the **p** side are very low! Indeed, they are even lower than in an intrinsic semiconductor (recall that **n** doping reduces the concentration of holes, whereas **p** doping reduces the concentration of free electrons). Consequently, the net drift current arising from reverse bias is very low (recall that the drift current is proportional to the concentration of charges – see Equation (1.77)). Thus, we conclude that the **p-n** junction is a very poor conductor in reverse bias (see Box 10).

Box 10   Gravitational analogue of the p-n junction in equilibrium, in forward bias and in reverse bias.

The physics of **p-n** junctions can be nicely illustrated by means of a gravitational analogue. Picture a bunch of balls (playing the role of electrons) rolling randomly around the bottom side of a slope, as represented in Figure 10.1a. Furthermore, imagine that the balls have sufficient energy to go all the way up the slope, but not sufficient to cross it, so they may reach the top, but end up coming back down (ignore collisions between the balls – in the 2D drawing, the balls going up obviously would collide with the ones coming down, but the system is actually 3D so they don't need to collide).

The system of Figure B10.1a is analogous to a **p-n** junction in equilibrium: the slope height is just enough to prevent the balls from reaching the other side, in the same way the internal electrostatic potential difference is just enough to cancel the chemical potential difference. That means that all balls going up also come down, and the net flux is zero. The balls going up are the diffusion current and the balls coming down are the drift current (remember that, in the real system, the electric current has the opposite direction of the flux of electrons, but in our gravitational analogue the balls have no charge, so this effect is not included in the analogy). Since all balls going up also come down, the drift and diffusion currents cancel each other.

Now suppose that the slope height is reduced, as represented in Figure B10.1b. Now the balls can easily flow to the other side, thus constituting a high current. This system is analogous to the **p-n** junction in forward bias. The system of Figure B10.1c, on the other hand, is analogous to the **p-n** junction in reverse bias. Now the slope is even higher than in Figure B10.1a, so, obviously, all balls going up keep coming down, and the net current is zero. Any "extra" component to the current would require "extra" balls on the top of the slope to come down, but there are no balls on the top (which is analogous to the very low concentration of electrons in the **p** region).

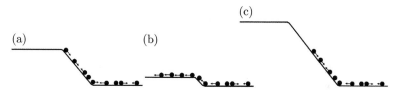

Figure B10.1 gravitational analogue of the physics of p-n junctions. a) analogue of the p-n junction in equilibrium. There is a high concentration of balls (the electrons) in the lower part, and the balls are rolling about randomly (thermal energy). The slope height (the electrostatic potential difference) is just enough to prevent the balls from reaching the top side (the p side). The balls going uphill constitute the diffusion current, and the balls going downhill constitute the drift current. Since no ball reaches the top, all balls going up also comes down, and the net current is zero. b) forward biased: the slope has been reduced, so now the balls can easily flow to the other side, constituting a high current. c) reverse biased: the slope was increased, so all balls going up also come back down, and the net current is zero. Any increase in the drift current would require "extra" balls from the top side to fall down, but there are no extra balls on the top side (very low concentration of electrons in the p region).

This asymmetrical conductivity of the **p-n** junction gives its rectifying property, which is so important in electronics. For example, circuits that convert oscillating signals (AC) to continuous signals (DC) rely on the rectifying property of diodes (see Box 11). Recall that a diode is just a **p-n** junction. The symbol for diodes that are used in circuits is shown in Figure 3.7.

In the previous section, we learned that the signature of a semiconductor out of equilibrium is the splitting of the Fermi level into *quasi*-Fermi levels. We noticed that this splitting always happens no matter how equilibrium is broken. So, it is not surprising that the Fermi level also splits in a biased **p-n** junction. The splitting is illustrated in the energy band diagrams of Figure 3.8. To explain the main characteristics of the band diagram and

Box 11   Diodes as rectifiers

A quintessential application of diodes is rectification, which is a process required to convert an alternating voltage (or AC voltage) into a constant voltage (or direct – DC voltage). For example, power is transmitted to our houses in AC form, so the voltage available in our walls is AC. But, to use it, we often need to convert it into DC, which requires rectification.

First, let me give you two examples of rectification, and then I will show you the simplest circuit that does the job. Figure B11.1 illustrates the two main types of rectification: the halfwave rectification and the full wave rectification. The top plot illustrates an incoming alternating voltage (a sinusoidal signal). In the half wave rectification, the negative part of the signal is chopped off, while in the full wave rectification the negative part is flipped into the positive part.

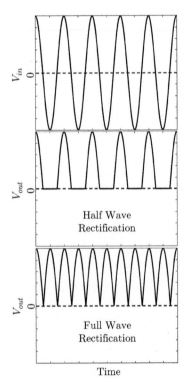

Figure B11.1 Illustration of rectification. The top figure illustrates an incoming sinusoidal wave. The middle figure illustrates halve wave rectification, wherein the negative part of the cycle is chopped off. The bottom figure illustrates full wave rectification, wherein the negative part of the cycle is flipped into the positive part.

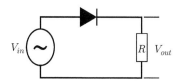

Figure B11.2 circuit of a half wave rectifier. In the positive cycle, the diode is forward biased and behaves essentially as a short-circuit. Consequently, the source voltage appears across the load R. In the negative cycle, the diode is reverse biased and behaves as an open circuit. Consequently, no current flows through the circuit and the voltage across the load is zero. In this case, the source voltage appears across the reverse biased diode.

Both types of rectifications use diodes, but the half wave can be done with only one diode, while the full wave requires at least four. Since the idea is essentially the same in both cases, I show only the circuit for half wave rectification in Figure B11.2. The circuit is very simple: it involves a diode and a load (a resistor), across which the output voltage is taken. In the positive cycle of the source, the diode is forward biased, so it behaves essentially as a short-circuit, thus allowing a current to flow and the transmission of the source voltage to the load. In the negative cycle, on the other hand, the incoming voltage reverse bias the diode, which now behaves as an open circuit. Thus, no current flows in the negative cycle, so no voltage appears across the load (the source voltage appears across the reverse biased diode).

To complete the job of turning an alternating voltage into a constant voltage, one can use a capacitor in parallel with the load, as shown in Figure B11.3a. In the positive cycle, the capacitor withholds a high voltage on the **n** side of the diode, thus reverse biasing it. Once the diode is reverse biased, the capacitor and load are isolated from the source. As shown in Figure B11.3b, the capacitor then discharges onto the load, so that the voltage drops almost linearly (actually, it is an exponential decay), until the source voltage is sufficiently high to forward bias the diode again, thus allowing the capacitor to be recharged by the source, and the cycle repeats. The higher is the product RC, the slower is the capacitor discharge. So, by choosing a capacitance and resistance sufficiently high, it is possible to make the discharge so slow that the line is nearly horizontal, thus resulting in a constant voltage. As a reference, the dashed curve in Figure B11.3 shows the positive cycle of the source (or, equivalently, the output voltage without the capacitor).

(Continued)

(Continued)

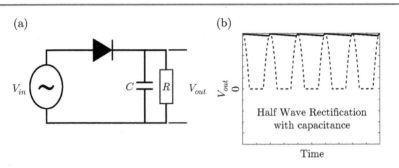

Figure B11.3 a) rectifier with capacitor. b) output voltage. In the positive cycle, the capacitor withholds a high voltage on the n side of the diode, thus reverse biasing it. Once the diode is reverse biased, the capacitor and load are isolated from the source. The capacitor then discharges its voltage onto the load, resulting in the almost linear decay shown in b). When the source voltage is sufficiently high to forward bias the diode again, the capacitor is recharged and the cycle repeats. The dashed line shows the positive cycle of the source (or, equivalently, the output voltage without the capacitor).

the Fermi level splitting, I will focus attention on the forward bias condition, which is sufficient to understand the band diagram for reverse bias as well.

Figure 3.8 shows the energy band diagram of the junction in equilibrium (no bias – top figure), forward biased (middle figure) and reverse biased (bottom figure). Focusing attention on the forward bias configuration (middle figure), we noticed earlier that the electrostatic potential barrier is lowered when the junction is forward biased. Since it is precisely the electrostatic potential difference that bends the energy diagram, the lower potential barrier results in a smoothing of the bend. How much the barrier is lowered (and the bending is smoothed) depends on the external potential difference, which we denote by V_{ext}. It is this external potential difference that injects the "extra" energy, which pushes the system away from thermodynamic equilibrium. Thus, the separation of the *quasi*-Fermi levels is also determined by V_{ext}. Following the notation of the previous sections, we denote the *quasi*-Fermi level for electrons as ε_{QFN}, and the *quasi*-Fermi level for holes as ε_{QFP}, see Figure 3.8.

The behaviour of the *quasi*-Fermi levels can be understood by inspecting the concentration of free charges. Consider the concentration of free electrons. According to Equation (3.32), the concentration of free electrons depends on the "distance" between ε_{QFN} and ε_C, that is, on $\varepsilon_C - \varepsilon_{QFN}$. Since the concentration of free electrons in the **n** region is already high, there is not

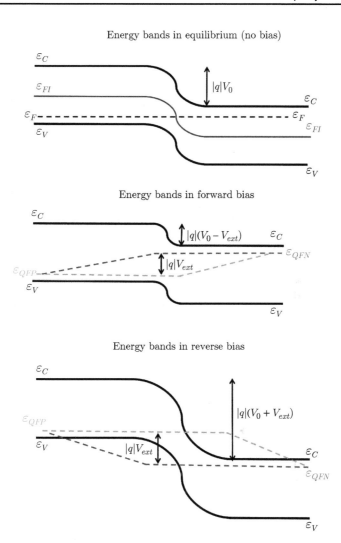

Figure 3.8 Energy band diagram without bias (top), with forward bias (middle) and with reverse bias (bottom).

much of a difference between its value within and without equilibrium. Thus, in the **n** side, the difference $\varepsilon_C - \varepsilon_{QFN}$ out of equilibrium is virtually identical to the difference $\varepsilon_C - \varepsilon_F$ in equilibrium ($\varepsilon_C - \varepsilon_{QFN} \approx \varepsilon_C - \varepsilon_F$ in the **n** side – compare Figure 3.8 top with Figure 3.8 middle). Inside the depletion region, there is a slight reduction in ε_{QFN} due to recombination, but usually the recombination process is sufficiently low for its effect to be ignored in a first approximation. Thus, ε_{QFN} is nearly constant across the depletion region. Things get interesting, however, when we consider the behaviour of ε_{QFN}

in the **p** side. We know that, in equilibrium, the concentration of free electrons is very low in the **p** side. Thus, the "extra" electrons arriving in the **p** side when the potential barrier is lowered changes the concentration dramatically. This change is manifested as a much shorter distance $\varepsilon_C - \varepsilon_{QFN}$ at the interface between the **p** side and the depletion region, when compared to $\varepsilon_C - \varepsilon_F$ in equilibrium ($\varepsilon_C - \varepsilon_{QFN} \ll \varepsilon_C - \varepsilon_F$ at the interface between the **p** side and the depletion region, compare the middle and top diagrams of Figure 3.8). These extra free electrons arriving in the **p** side, however, diffuse in a region with a high concentration of holes. Thus, some of the free electrons recombine with the holes, that is, they fall back from the conduction band to the valence band, thus annihilating a pair of free charges. Consequently, the concentration of free electrons lowers across the **p** side, until it reaches its equilibrium value. Such lowering in the concentration is manifested as a gradually larger separation $\varepsilon_C - \varepsilon_{QFN}$, until it reaches the equilibrium condition ($\varepsilon_C - \varepsilon_{QFN} \approx \varepsilon_C - \varepsilon_F$ in the **p** side far away from the depletion region). Thus, the condition $\varepsilon_C - \varepsilon_{QFN} \approx \varepsilon_C - \varepsilon_F$ is observed in two different regions for two different reasons: it is observed in the **n** side due to the high concentration of free electrons, and it is observed in the **p** side away from the depletion region because no "extra" charges reaches this region.

An analogous reasoning accounts for the behaviour of ε_{QFP}: since the concentration of holes is high in the **p** side, there is not a significant difference in this concentration within and without equilibrium, so $\varepsilon_{QFP} - \varepsilon_V \approx \varepsilon_F - \varepsilon_V$ in the **p** region. However, the concentration of holes at the interface between the depletion region and the **n** side is much higher for forward bias than in equilibrium, thus $\varepsilon_{QFP} - \varepsilon_V \ll \varepsilon_F - \varepsilon_V$ at the interface between the depletion region and the **n** side. Away from the interface, the concentration of holes drops due to recombination, until it reaches its value in equilibrium.

As shown in Figure 3.8, the separation of the *quasi*-Fermi levels is set by the external voltage, that is:

$$|q| V_{ext} = \varepsilon_{QFN} - \varepsilon_{QFP} \quad for \quad -x_p \leq x \leq x_n \tag{3.34}$$

It is instructive to reflect on why Equation (3.34) must be true. What does it mean to assert that there is an external voltage V_{ext} across the **p-n** junction? It means that the charge carriers lose an energy of $|q| V_{ext}$ as they travel through the device (notice that this is the same interpretation for the voltage drop in a resistor). In forward bias, the current enters the junction in the **p** side and leaves from the **n** side. That means that electrons are entering in the **n** side and leaving from the **p** side. So, to assert that there is a voltage drop of V_{ext} in forward biasing is tantamount to asserting that the energy of the electrons in the extreme of the **n** side (the "end" of the **p-n** junction) is higher than the energy at the beginning of the **p** side (the "beginning" of the **p-n** junction) by the amount $|q| V_{ext}$. More specifically, we have seen in Chapter 1 that this energy difference is a difference in the electrochemical energy, that is, the sum of the chemical and electrostatic potentials. In other words, it is a difference in the *quasi*-Fermi level ε_{QFN} between both sides of the junction. Thus, to assert that there is a voltage drop of V_{ext} across the junction in forward bias is tantamount to the assertion that ε_{QFN} in the extreme of the **n** side (end of the **p-n** junction) is higher than ε_{QFN} at the beginning of the **p** side (beginning of the **p-n** junction) by $|q| V_{ext}$. But ε_{QFN} at the beginning of the **p** side (which is far away from the depletion region) coincides with ε_{QFP} (see Figure 3.8, middle). Consequently, the assertion that "ε_{QFN} in the **n** side is higher than ε_{QFN} in the **p** side by $|q| V_{ext}$" is logically equivalent to the assertion that $\varepsilon_{QFN} - \varepsilon_{QFP} = |q| V_{ext}$, when ε_{QFN} is taken in the extreme of the **n** side and ε_{QFP} is taken at the beginning of the **p** side. This latter assertion is, of course, the assertion of Equation (3.34) (notice that the same conclusion could be reached by considering the energy of holes). Thus, if you come to think of it, Equation (3.34) is a tautology: it is asserting that the external voltage (left-side of the equality) is the external voltage (right-side of the equality).

It is convenient to classify the charge carriers into "minority carriers" and "majority carriers". Free electrons in the **n** side and holes in the **p** side are majority carriers, whereas free electrons in the **p** side and holes in the **n** side are minority carriers. Inside the depletion region, the concentrations of both carriers are not very different, so it does not make sense to distinguish them between minority and majority carriers.

The total current that flows in a **p-n** junction is the combination of the current of free electrons and the current of holes, as shown in Figure 3.9. In the **n** side, far away from the depletion region, the concentration of holes is essentially the same as in equilibrium (there are no "extra" holes). Therefore, the current far away from the depletion region in the **n** side is due solely to free electrons. For analogous reasons, the current in the **p** side far away from the depletion region is due solely to holes. Using the classification of carriers, we can combine these two facts into one single assertion: "the current far away from the depletion region is due to majority carriers". Near the depletion region, however, the contribution of minority carriers is significant, so the total current is due to both minority and majority carriers. Furthermore, the current within the depletion region is almost constant, again due to low recombination rates. The fact that the currents do not change significantly inside the depletion region allows the calculation of the total current to be performed only in terms of the minority carriers, which greatly facilitates this task. This is, however, the subject of the next section.

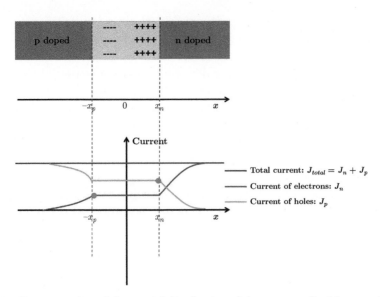

Figure 3.9 Representation of the spatial distribution of the currents. In this example, the current of holes in the depletion region is higher than the current of electrons, but this is not necessarily the case. Depending on how the semiconductor is doped, the current of electrons could be higher, or they could also be the same. The orange dots indicate the positions where the currents are calculated.

3.5   The current vs voltage relationship in a p-n junction (Shockley equation)

We have reached the crowning section of this book. Relying on the conceptual tools developed previously, here we will derive the relationship between current and voltage in a **p-n** junction.

Our task is to calculate the total current, which is shown as the blue line in Figure 3.9. The total current must be constant across the **p-n** junction, otherwise there would be accumulation of charges (since the current that enters the diode is the same as the current that exits the diode, there can be no accumulation of charges inside the diode). Thus, we could choose any point in Figure 3.9 to calculate the total current. A sound strategy, therefore, is to begin by identifying the point that most facilitates the calculation.

Following this strategy, let us make a survey of the characteristics of the different regions across the **p-n** junction. Let us begin by considering the **n** side, that is, the region $x \geq x_n$ in Figure 3.9. We already know that there are two types of charge carriers: minority carriers and majority carriers. But what types of currents do we have in the **n** side? To answer this question, first notice that the electric field is confined within the depletion region, so there is no electric field in the **n** side (nor in the **p** side). To be sure, this is just an approximation, but a very good one. Why is the electric field almost fully confined within the depletion region? The answer is: for the same reason that the electric field in metals is nearly null: due to the high conductivity of the **n** and **p** sides. This is an important point that is often overlooked in the literature, so let us discuss it in more detail.

We know that there are two types of currents: drift and diffusion. As seen in section 3.2, the drift current is proportional to the electric field: $J_{drift} = \rho \cdot \langle v \rangle = \rho \cdot \mu \cdot E$. The proportionality between the electric field and the drift current is defined as the conductivity σ of the material. Thus, the drift current of majority carriers in the **n** side can be expressed as:

$$J_{drift_n} = \sigma_n^{n\ side} \cdot E \qquad for\ x \geq x_n \qquad (3.35)$$

The notation used in Equation (3.35) deserves some spelling out. The downstairs index "*drift n*" indicates that $J_{drift\,n}$ is the drift current of electrons (which are the majority carriers in the **n** side). The upstairs index "**n side**" indicates that the conductivity $\sigma_n^{n\ side}$ is the conductivity in the **n** side. Finally, the downstairs index "*n*" indicates that this is the conductivity of electrons. Thus, $\sigma_n^{n\ side}$ is the conductivity of electrons in the **n** side.

The high concentration of free electrons in the **n** side results in a high $\sigma_n^{n\ side}$. Thus, even though $J_{drift\,n}$ is a significant contribution to the total current, the electric field E is very small in the **n** region due to the high $\sigma_n^{n\ side}$ (in the limit that $\sigma_n^{n\ side}$ goes to infinity, E must go to zero to keep $J_{drift\,n}$ finite). In other words, it takes a tiny E to set up a high $J_{drift\,n}$ in the **n** side. In the **p** side, it is the high concentration of holes that accounts for the vanishingly small electric field. Since the field is very weak in both **n** and **p** sides, it is nearly fully confined within the depletion region.

There is also a contribution of diffusion current of majority carriers. In the **n** side, this contribution is:

$$J_{dif_n} = |q| D_N \frac{\partial n}{\partial x} \qquad for\ x \geq x_n \qquad (3.36)$$

where D_N is the electron diffusion coefficient.

Now let us turn attention to the current of minority carriers. In the **n** side, the drift current of minority carriers (holes) is:

$$J_{drift_p} = \sigma_p^{n\ side} \cdot E \qquad for\ x \geq x_n \qquad (3.37)$$

The term $\sigma_p^{n\ side}$ is the conductivity of holes in the **n** side. But the concentration of holes in the **n** side is very low, so $\sigma_p^{n\ side}$ is also very low.

Furthermore, we had already concluded that E is vanishingly small outside the depletion region. Thus, the current of minority carriers is given by the product of two vanishingly small quantities, so it is also a vanishingly small quantity. Consequently, we can dismiss the drift current of holes in the **n** side as insignificant, that is:

$$J_{drift_p} = \sigma_p^{\text{n side}} \cdot E \approx 0 \quad for \ x \geq x_n \tag{3.38}$$

Analogous considerations can be made for minority carriers in the **p** side, which leads to the conclusion that:

$$J_{drift_n} = \sigma_n^{\text{p side}} \cdot E \approx 0 \ for \ x \leq -x_p \tag{3.39}$$

Notice that $J_{drift\ n}$ in Equation (3.39) is the drift current of electrons in the **p** side, whereas the $J_{drift\ n}$ in the Equation (3.35) is the drift current of electrons in the **n** side. So, they are not the same quantity.

Summarizing the contributions of each type of carrier, we have:

$$\textbf{n side } (x \geq x_n)$$

$$J_n(x) = J_{drift_n} + J_{dif_n} = \sigma_n^{\text{n side}} \cdot E + |q|D_N \frac{\partial n}{\partial x} \ (1)$$

$$J_p(x) = J_{dif_p} = -|q|D_P \frac{\partial p}{\partial x} \ (2) \tag{3.40}$$

$$\textbf{p side } \left(x \leq -x_p \right)$$

$$J_n(x) = J_{dif_n} = |q|D_N \frac{\partial n}{\partial x} \ (1)$$

$$J_p(x) = J_{drift_p} + J_{dif_p} = \sigma_p^{\text{p side}} \cdot E - |q|D_P \frac{\partial p}{\partial x} \ (2) \tag{3.41}$$

where $J_n(x)$ and $J_p(x)$ are the currents of free electrons and holes, respectively.

We have reached the important conclusion that the only significant contribution for the current of minority carriers is the diffusion current. Can we take advantage of this insight to select the point where we are going to evaluate the total current? Ideally, we would like to select a point where the contribution of minority carriers dominates, so that we could deal only with diffusion currents. Unfortunately, there is no point where only minority carriers contribute, as can be seen from inspection of Figure 3.9: wherever a current of minority carrier is significant, there is also a significant current of majority carriers. Hang on, though. What if we calculated the current of minority carriers at the edges of the depletion region? If we did that, we could take advantage of the fact that the current is constant across the depletion region, and then obtain the total current by working out the diffusion current of minority carriers on both edges!! More specifically, we can calculate the current of holes at the point x_n and the current of free electrons at the point $-x_p$ (see the orange dots in Figure 3.9). Both of these are minority carrier currents, so they are both diffusion currents. Since the currents are constant across the depletion region, we can add both contributions to find the total current, that is:

$$J_{total} = J_p(x_n) + J_n(-x_p) \qquad (3.42)$$

Our task has thus been reduced to: 1) find the concentrations of minority carriers, 2) use the concentrations to find the diffusion currents at the edges of the depletion region, 3) add them up to obtain the total current.

To find the concentration of minority carriers, notice that their transport is by diffusion and that there is a loss mechanism (the recombination process). How does the concentration of charge carriers behave in this condition? We have already examined this problem in section 1.11, where we concluded that the concentration of charge carriers decays exponentially when transport is by diffusion in the presence of losses (Equation (1.110) and Equation (1.111)). For the sake of convenience, I repeat Equation (1.110) below, but assuming an x-dependence, as in Figure 3.9:

$$p(x) = p(0) \cdot \exp\left(-\frac{x}{L_P}\right)$$

There is a subtlety, though: the loss mechanism applies only to the "extra" charge carriers. To be sure, both the equilibrium charges and the "extra" charges due to the current flow suffer recombination, but the recombination of the former is compensated by the opposite effect: generation due to thermal excitation of electrons to the conduction band (in thermal equilibrium, there is an ongoing recombination and generation process, but they cancel each other, so that the concentrations do not change). Thus, there is no net loss mechanism for the charges in equilibrium. I will describe this fine balance in more detail in a minute, but first let us define a bit of notation: the concentration of holes in thermodynamic equilibrium will be denoted by p_0 and the concentration of "extra" holes will be denoted by p_e. We still designate the total concentration of holes as p, so:

$$p = p_0 + p_e \qquad (3.43)$$

As the loss mechanism applies only to the excess concentration p_e, so Equation (1.110) applies only to p_e. Adapting this equation to the coordinate system of Figure 3.9, we find:

$$p_e(x) = p_e(x_n) \cdot \exp\left(-\frac{x-x_n}{L_P}\right) \quad for \ x \geq x_n \tag{3.44}$$

An analogous consideration leads to an expression for the minority carriers in the **p** side. Thus:

$$n_e(x) = n_e(-x_p) \cdot \exp\left(\frac{x+x_p}{L_N}\right) \quad for \ x \leq -x_p \tag{3.45}$$

where n_e is the excess of free electrons.

Equation (3.44) and Equation (3.45) can now be used to evaluate the current densities at the edges of the depletion region. Before we do so, however, it is apposite to explain the dynamics of recombination and generation in a bit more detail. I will show it for the concentration of holes, but, as usual, the same considerations also apply to free electrons.

According to Equation (1.103), the relationship between current and hole concentration is given by:

$$-\frac{\partial J}{\partial x} - |q|\frac{p}{\tau_p} = |q|\frac{\partial p}{\partial t}$$

Recall that τ_p is the recombination lifetime. Thus, the second term of the equation is the recombination rate, which we designate as R:

$$R = |q|\frac{p}{\tau_p}$$

If there is also a process of generation of holes (in this case, by thermal excitation), then we need to include it in the equation as well. Denoting the generation rate as G, we then get:

$$-\frac{\partial J}{\partial x} - R + G = |q|\frac{\partial p}{\partial t} \tag{3.46}$$

We are interested in the situation where J is a diffusion current, so Equation (3.46) reduces to:

$$-|q|D_P\frac{\partial^2 p}{\partial x^2} - R + G = |q|\frac{\partial p}{\partial t} \tag{3.47}$$

The temporal derivative vanishes in stationary regime. Thus:

$$|q| D_P \frac{\partial^2 p}{\partial x^2} = R - G \qquad (3.48)$$

In thermodynamic equilibrium $p = p_0$, so:

$$|q| D_P \frac{\partial^2 p_0}{\partial x^2} = |q| \frac{p_0}{\tau_P} - G \qquad (3.49)$$

Notice that we expressed R in terms of the lifetime in Equation (3.49). In equilibrium, the concentrations within either the **p** or the **n** side cannot vary (otherwise, there would be a diffusion current), so:

$$\frac{\partial^2 p_0}{\partial x^2} = 0$$

From which we conclude that:

$$G = |q| \frac{p_0}{\tau_P} \qquad (3.50)$$

Equation (3.50) gives an expression for the thermal generation of charge carriers. It is important to notice that it only applies to generation by thermal excitation (since we derived it by invoking thermodynamic equilibrium). It does not apply when there is a different source of generation, like illumination, for example. However, here, we are interested only in thermal excitation. Substituting it into Equation (3.46), we find:

$$|q| D_p \frac{\partial^2 p}{\partial x^2} = R - G = |q| \frac{p}{\tau_P} - |q| \frac{p_0}{\tau_P} \qquad (3.51)$$

Expressing the total concentration in terms of the concentrations in equilibrium and the "extra" concentration, Equation (3.51) reads:

$$|q| D_P \frac{\partial^2 (p_e + p_0)}{\partial x^2} = |q| \frac{p_e + p_0}{\tau_P} - |q| \frac{p_0}{\tau_P} = |q| \frac{p_e}{\tau_P}$$

but:

$$\frac{\partial^2 p_0}{\partial x^2} = 0$$

so:

$$D_P \frac{\partial^2 p_e}{\partial x^2} = \frac{p_e}{\tau_P} \qquad (3.52)$$

This is the same relation we found in section 1.11 (Equation (1.107)) but applied to the excess of charge carriers. Equation (3.44) is the solution to Equation (3.52) in the coordinate system of Figure 3.9.

Returning to our main task, now we need to evaluate the diffusion currents of minority carriers. Thus, substituting Equation (3.44) into Equation (3.40):

$$J_p(x) = J_{dif_p}(x) = -|q|D_P \frac{\partial(p_0 + p_e)}{\partial x} = -|q|D_P \frac{\partial p_e}{\partial x} = -|q|D_P p_e(x_n)$$

$$\cdot \frac{\partial \exp\left(-\frac{x - x_n}{L_P}\right)}{\partial x}$$

$$\therefore J_p(x) = \frac{|q|D_P}{L_P} \cdot p_e(x_n) \cdot \exp\left(-\frac{x - x_n}{L_P}\right) \quad \text{for } x \geq x_n \tag{3.53}$$

Similarly, substituting Equation (3.45) into Equation (3.41):

$$J_n(x) = J_{dif_n}(x) = |q|D_N \frac{\partial(n_0 + n_e)}{\partial x} = |q|D_N \frac{\partial n_e}{\partial x} = |q|D_N n_e(-x_p)$$

$$\cdot \frac{\partial \exp\left(\frac{x + x_p}{L_N}\right)}{\partial x}$$

$$\therefore J_n(x) = \frac{|q|D_N}{L_N} \cdot n_e(-x_p) \cdot \exp\left(\frac{x + x_p}{L_N}\right) \quad \text{for } x \leq -x_p \tag{3.54}$$

Finally, substituting Equation (3.53) and Equation (3.54) back into Equation (3.42):

$$J_{total} = J_p(x_n) + J_n(-x_p) = \frac{|q|D_P}{L_P} \cdot p_e(x_n) + \frac{|q|D_N}{L_N} \cdot n_e(-x_p) \tag{3.55}$$

That's it: we have an expression for the current flowing through a diode. But if you gift an electrical engineer with Equation (3.55), you are likely to receive an indignant slap on the face for a thank you. Alas, we live in a world of spoiled electrical engineers, who insist that currents be expressed in terms of voltages. But where is the voltage in Equation (3.55)? Well, you may have guessed: it is embedded in the terms $p_e(x_n)$ and $n_e(-x_p)$ (notice that both terms must be zero if there is no external voltage – in thermal equilibrium $p = p_0$ and $n = n_0$). So, we need to express these terms explicitly in terms of the voltage to get an acceptable expression for the current.

We begin with $n_e(-x_p)$. This is the extra concentration of electrons at the interface between the **p** side and depletion region. From Equation (3.32), the total concentration at the interface is:

$$n\left(-x_p\right) = n_i \exp\left[-\frac{\varepsilon_{FI}\left(-x_p\right) - \varepsilon_{QFN}\left(-x_p\right)}{k_B T}\right] \tag{3.56}$$

On the other hand, according to Equation (3.33), the concentration of holes at the same interface is:

$$p\left(-x_p\right) = p_i \exp\left[-\frac{\varepsilon_{QFP}\left(-x_p\right) - \varepsilon_{FI}\left(-x_p\right)}{k_B T}\right] \tag{3.57}$$

Taking the product between these two concentrations:

$$n\left(-x_p\right) \cdot p\left(-x_p\right) = n_i \cdot p_i \exp\left[\frac{\varepsilon_{QFN}\left(-x_p\right) - \varepsilon_{QFP}\left(-x_p\right)}{k_B T}\right] \tag{3.58}$$

With the help of Equation (3.34), Equation (3.58) can be recast as:

$$n\left(-x_p\right) \cdot p\left(-x_p\right) = n_i \cdot p_i \exp\left(\frac{|q| V_{ext}}{k_B T}\right) \tag{3.59}$$

We can express Equation (3.59) in terms of the concentrations in equilibrium. Recall that the product $n \cdot p$ is constant in equilibrium, that is: $n_i \cdot p_i = n_0 \cdot p_0$. Thus, Equation (3.59) can be written as:

$$n\left(-x_p\right) \cdot p\left(-x_p\right) = n_0\left(-x_p\right) \cdot p_0\left(-x_p\right) \exp\left(\frac{|q| V_{ext}}{k_B T}\right) \tag{3.60}$$

According to Equation (3.60), the actual product $n \cdot p$ is higher than the product in equilibrium ($n_0 \cdot p_0$) by the factor $\exp\left(\frac{|q| V_{ext}}{k_B T}\right)$, which makes sense, since the system is not in equilibrium.

Equation (3.60) involves the product between majority carriers (holes) and minority carriers (free electrons). As discussed in the previous section, the concentration of majority carriers is not significantly altered by the application of an external voltage, that is:

$$p\left(-x_p\right) \approx p_0\left(-x_p\right) \tag{3.61}$$

Thus, with the help of Equation (3.61), Equation (3.60) is reduced to:

$$n\left(-x_p\right) = n_0\left(-x_p\right) \exp\left(\frac{|q| V_{ext}}{k_B T}\right) \tag{3.62}$$

But $n(-x_p) = n_e(-x_p) + n_0(-x_p)$, so:

$$n_e\left(-x_p\right) = n\left(-x_p\right) - n_0\left(-x_p\right)$$

Therefore:

$$n_e\left(-x_p\right) = n_0\left(-x_p\right)\left[\exp\left(\frac{|\,q\,|\,V_{ext}}{k_B T}\right) - 1\right] \qquad (3.63)$$

Notice that $n_0(-x_p)$ is the equilibrium concentration of free electrons in the **p** side. In Chapter 2, we learned how to determine this concentration from the concentration of dopants. So, Equation (3.63) expresses $n_e(-x_p)$ in terms of the concentration of dopants (through $n_0(-x_p)$) and the external voltage.

An analogous procedure leads to the concentration of holes at the interface between the depletion region and the **n** side:

$$p_e(x_n) = p_0(x_n)\left[\exp\left(\frac{|\,q\,|\,V_{ext}}{k_B T}\right) - 1\right] \qquad (3.64)$$

Substituting Equation (3.64) and Equation (3.63) back into Equation (3.55), we find:

$$J_{total} = |\,q\,|\left[\frac{D_P}{L_P}p_0(x_n) + \frac{D_N}{L_N}n_0\left(-x_p\right)\right]\left[\exp\left(\frac{|\,q\,|\,V_{ext}}{k_B T}\right) - 1\right] \qquad (3.65)$$

In the context of electronic circuits, it is more useful to know the total current I instead of the current density J. Thus, denoting the cross-sectional area of the **p-n** junction by A, the current is given by:

$$I = \left[A|q|\frac{D_P}{L_P}p_0(x_n) + A|q|\frac{D_N}{L_N}n_0\left(-x_p\right)\right]\left[\exp\left(\frac{|\,q\,|\,V_{ext}}{k_B T}\right) - 1\right] \qquad (3.66)$$

Notice that the terms between brackets depend on parameters of the semiconductor (through A, D_N, D_P, L_N and L_P), and on the concentration of dopants (through $p_0(x_n)$ and $n_0(-x_p)$). This term is the saturation current, usually denoted by I_0 or I_S. Adopting the former notation:

$$I_0 = A\,|\,q\,|\left[\frac{D_P}{L_P}p_0(x_n) + \frac{D_N}{L_N}n_0\left(-x_p\right)\right] \qquad (3.67)$$

Therefore:

$$I = I_0\left[\exp\left(\frac{|\,q\,|\,V_{ext}}{k_B T}\right) - 1\right] \qquad (3.68)$$

Equation (3.68) is the magnificent Shockley equation, first derived by the ubiquitous William B. Shockley Jr, arguably the person who most contributed to the development of electronics in the twentieth century.

Shockley equation expresses the current in terms of the external voltage in a **p-n** junction. Even though we derived the equation assuming forward bias ($V_{ext} > 0$), all the assumptions also apply to reverse bias ($V_{ext} < 0$). Therefore, Shockley equation is valid for both biasing conditions. Notice that, for reverse bias ($V_{ext} < 0$), the exponential acquires values much smaller than one for modest negative V_{ext}. Thus, for negative bias, the current is nearly equal to the saturation current:

$$I \sim I_0 \ for \ V_{ext} < 0$$

The saturation current, however, is typically very small. In silicon diodes, for example, the saturation current is of the order of 10^{-9} A. That means that a **p-n** junction is a very poor conductor when reverse biased, as we had already concluded in the previous section. For forward bias, on the other hand, the current increases exponentially with the voltage, which means that the **p-n** junction is an excellent conductor when forward biased. These features are nicely captured by an IxV plot of Shockley equation (Equation (3.68), Equation (3.69)), as shown in the blue line of Figure 3.10, which assumes $I_0 = 10^{-9}A$. The parameter ς, which I sneaked into the plot, is known as the ideality factor. This parameter captures some effects that we did not take into account in our derivation, mainly recombination in the depletion region. It takes on values between one and two, and it enters Shockley equation as:

$$I = I_0 \left[\exp \left(\frac{\mid q \mid V_{ext}}{\varsigma k_B T} \right) - 1 \right] \tag{3.69}$$

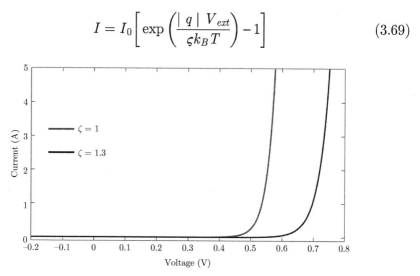

Figure 3.10 Current x Voltage plot of Shockley equation assuming $I_0 = 10^{-9}A$ and for two conditions of ideality factor ς.

I added a plot of a second curve for a higher value of ς, just for you to get a feel of what it does, but a full treatment of the ideality factor is beyond the scope of this book (see section 3.6 for suggestions on more advanced books).

If you take a course in electronic circuits, one of the first things you will learn is how to analyse circuits involving diodes. Compared to the analysis of circuits involving only resistors, capacitors and inductors, the presence of diodes complicates the analysis because Shockley equation is non-linear. But, as a first approximation, one can treat the diode as an open circuit if it is operating in reverse bias, and as short-circuit if it is operating in forward bias. The reason why treating it as a short-circuit in forward bias is a good approximation can be understood by inspection of Figure 3.10: it takes a huge change in current to obtain a small change in voltage. Thus, in practice, the voltage is always about the same, not mattering the actual current that is flowing. If the voltage is small compared to the source, then you can just ignore it (thus, a short-circuit). If it is not, you may have to consider a constant voltage drop, but that is still much simpler than considering the full Shockley equation. In the examples of Figure 3.10, the operation voltages are about 0.5 V for the ideal diode ($\varsigma = 1$) and about 0.7 V for the non-ideal diode ($\varsigma = 1.3$).

Box 12   Essential ideas in the analysis of circuits involving diodes

To illustrate the essential ideas involved in the analysis of circuits with diodes, consider the circuit shown in Figure B12.1:

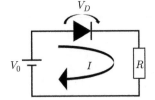

Figure B12.1 A circuit consisting of a forward biased diode in series with a resistor (load).

We will find the current I flowing through this circuit at two different conditions, one involving a low voltage source V_0 and low resistance R and then again assuming high voltage source V_0 and high resistance R. In all cases, I assume a saturation current $I_0 = 10^{-9}$ A.

First, let us lay down the equations governing the circuit dynamics. From Kirchhoff's voltage law, we get:

$$V_0 = V_D + I \cdot R$$

where V_D is the voltage drop across the diode and $I \cdot R$ is the voltage drop across the resistance R. It is more convenient to express this equation in terms of the current:

$$I = \frac{V_0 - V_D}{R}$$

Our goal is to find the current I when the source V_0 and resistance R have been specified. But we have two unknowns: the current I and the diode voltage V_D. If we have two unknowns, we need two equations. Of course, the other equation is Shockley equation, which relates I and V_D:

$$I = I_0 \left[\exp \left(\frac{|q| V_D}{k_B T} \right) - 1 \right]$$

Now we have two unknowns and two equations, and the problem is reduced to substituting one equation into the other and solving for either I or V_D.

Instead of solving these equations, it is more instructive to visualize them together in a plot of I against V_D. This kind of plot is very useful to give us an idea of the accuracy involved in the different approximations that can be used to find the current quickly, without having to solve the set of equations.

Such a plot is shown in Figure B12.2a, where the values $V_0 = 2\,V$ and $R = 10\,\Omega$ are assumed. The line coming from Kirchhoff's voltage law is called the "load line", as a reference to the linear dependence of current against voltage in a resistor (the load).

The first feature to be noticed in the plot of Figure B12.2a is that, when the diode is forward biased, the exponential dependence in Shockley equation entails that a tiny variation in the voltage V_D around a fixed value entails a huge variation in the current. The "fixed value" is the diode's operation voltage, and it is $\sim 0.5\,V$ in our example. Thus, instead of solving the two equations to find the exact values, one could find the current from Kirchhoff's law by ascribing the operation voltage to the diode. In our example, ascribing $V_D = 0.5\,V$, results in $I = 0.15\,A$. We can compare this result with the exact value, which, of course, is the point where the two curves cross. To get a feel for the differences one gets, a plot zoomed in the range $0.48 \leq V_D \leq 0.5$ is shown in Figure B12.2b. The exact value is $I = 0.1513\,A$, which is quite close to the approximated value of $I = 0.15\,A$.

(Continued)

(*Continued*)

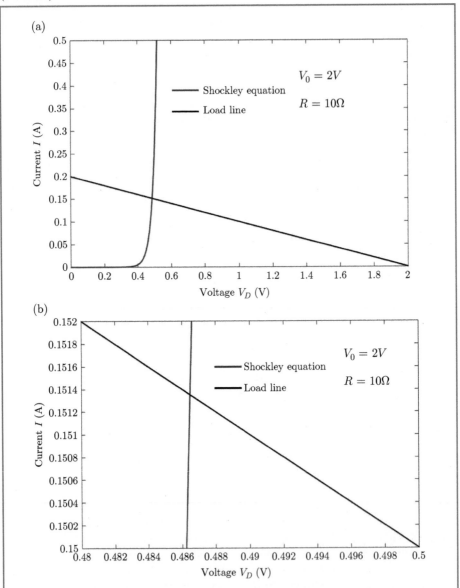

Figure B12.2 The vertical axis is the current I and the horizontal axis is the diode voltage V_D. a) load line plotted together with Shockley equation assuming $V_0 = 2\,V$ and $R = 10\,\Omega$ b) same as a), but zoomed in between the values $0.48 \leq V_D \leq 0.5$.

This approximation gets even better if a higher resistance is used. This is shown in Figure B12.3a), which assumes $V_0 = 200$ V and $R = 10$ $k\Omega$. Notice that the slope of the load line is given by the inverse of the resistance, so that a higher resistance entails a lower slope (that is, a more horizontal line). In this example, the line is almost horizontal within the range shown in the plot, so we cannot even see the point where the load line crosses the horizontal axis (of course, it crosses at $I = 0$, which entails $V_D = V_0 = 200$ V, that is, far, far away from $V_D = 0.5$ V). In this case, even treating the diode as a short-circuit current (that is, assuming $V_D = 0$) is a good approximation (see Figure B12.3b for a zoom around the exact value). The values obtained assuming $V_D = 0$, $V_D = 0.5$ and the exact values are shown in Table B12.1 for both examples.

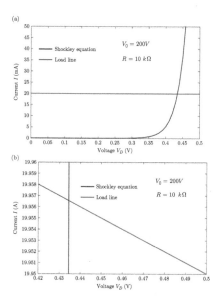

Figure B12.3 The vertical axis is the current I and the horizontal axis is the diode voltage V_D. a) load line plotted together with Shockley equation assuming $V_0 = 200$ V and $R = 10$ $k\Omega$ b) same as a), but zoomed in between the values $0.42 \leq V_D \leq 0.5$.

Table B12.1 Values obtained treating the diode as a short-circuit ($V_0 = 0\,V$), assuming a fixed operation voltage ($V_D = 0.5$), and the exact values.

	$V_0 = 2\,V$, $R = 10\Omega$	$V_0 = 200\,V$, $R = 10$ $k\Omega$
$V_D = 0$	$I \overset{.}{=} 0.2$ A	$I = 20$ mA
$V_D = 0.5$	$I = 0.15$ A	$I = 19.95$ mA
Exact	$I = 0.1513$ A	$I = 19.9565$ mA

We have accomplished the main purpose of this book, which is the understanding of the electrical properties of **p-n** junctions. In the next two chapters, I show the essential concepts of two key applications of **p-n** junctions: photovoltaic devices (solar cells and photodetectors) and transistors.

Box 13   What makes a diode a light emitting diode?

We have seen that there is an ongoing recombination process when a current flows through a diode. Since in the recombination process electrons fall from the conduction to the valence band, then recombination entails of loss of E_G in each electron (recall that E_G is the band gap energy). Where does this energy go?

The answer to this question depends on the so-called energy dispersion diagram. This diagram is a plot of the energy against the momentum of electrons in solids, and many important properties of the solids depend on the shape of these diagrams. Indeed, one major goal of semiconductor research is to find ways of engineering the dispersion diagrams to control their properties.

The energy dispersion diagrams motivate distinguishing semiconductors into two classes: direct band gap semiconductors and indirect band gap semiconductors. These two types of semiconductors are illustrated in a simplified manner in Figure B13.1 below.

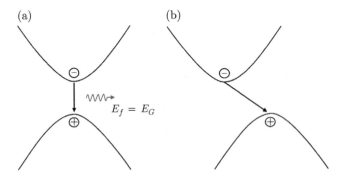

Figure B13.1 Simplified illustration of energy dispersion diagrams. The vertical axis is energy and the horizontal axis is momentum. a) direct band gap semiconductor b) indirect band gap semiconductor.

A direct band gap semiconductor is shown in Figure B.13.1a, and an indirect band gap semiconductor is shown in Figure B.13.1b. The vertical axis is energy and the horizontal axis is momentum. In a direct band gap semiconductor, the bottom of the conduction band is aligned with the top of the valence band, whereas is an indirect band gap semiconductor the edges of the bands are misaligned.

When an electron recombines with a hole in a direct band gap semiconductor, the momentum of the electron is not changed in the recombination process (the transition is vertical). This type of transition can be mediated by the emission of a photon. Thus, Light Emitting Diodes (LED) are made of direct band gap semiconductors. Since energy is conserved, the energy of the photon coincides with the energy of the band gap ($E_f = E_G$). Therefore, it is the band gap energy E_G that defines the colour of the LED (the energy of a photon is proportional to the frequency – $E_f = hf$ – and the frequency sets the colour). High band gap semiconductors emit high energy photons (which lie at the blue part of the spectrum), whereas low band gap semiconductors emit low-energy photons (which lie at the red part of the spectrum). The effect of light emission induced by an electric current is called "electroluminescence".

Recombination in an indirect band gap semiconductor, however, cannot be mediated by photons because not only energy, but also momentum, must be conserved. In an indirect band gap semiconductor, the momentum of the electron changes in the transition (the transition is diagonal). Therefore, the transition must be mediated by a particle having sufficiently high momentum to satisfy conservation. Photons have low momentum (they do not even have mass), so they cannot mediate this transition (notice that this is not a problem for direct band gap semiconductors because the momentum is not changed in the transition – it is a vertical transition). Thus, recombination in indirect band gap semiconductors is mediated by emission of phonons, which are quanta of vibration of the underlying atomic structure. These vibrations are associated with the temperature of the material, so the energy of the electrons is turned into heat.

Silicon is an indirect band gap semiconductor, so you will be hard-pressed to find a LED made of silicon (if you go on with your studies and end up finding a way to make silicon emit light efficiently, you may get quite rich and famous). Gallium arsenide is an example of an important direct band gap semiconductor used in the LED industry.

3.6   Suggestions for further reading

In this chapter, I tried to keep the notation and logical flow conversant with *Solid State Electronic Devices*, by Ben G. Streetman and Sanjay K. Banerjee, which I suggest as a second reading. For more advanced studies, the classic world reference is *Physics of Semicondutor Devices*, by S. M. Sze.

3.7 Exercises

Exercise 1
Argue that:

$$|q|E = \frac{\partial \varepsilon_{FI}}{\partial x}$$

Where q is the fundamental charge, E is the electric field within the depletion region of the **p-n** junction, whose intrinsic Fermi level is ε_{FI}.

Exercise 2
From the equilibrium condition in a **p-n**, junction, prove that

$$|q|D = \mu \cdot k_B \cdot T$$

Hint: start from the equilibrium condition for the currents, express the concentration of charge carriers in terms of the Fermi level and the intrinsic Fermi level, and use the equality of Exercise 1.

Exercise 3
Express the conductivity in terms of the concentration of charge carriers and the mobility.

Exercise 4
Find the resistance of a diode when $V_{ext} \gg \frac{k_B T}{|q|}$. Is the resistance at this condition high or low? Repeat the problem for $V_{ext} \ll \frac{k_B T}{|q|}$.

Hint: the resistance of a non-linear element is given by the derivative of the voltage with respect to the current. It is easier to find the inverse of the resistance than the resistance itself.

Exercise 5
Explain qualitatively the physical origin of the rectifying property of diodes.

Exercise 6
Sketch the energy band diagram of a **p-n** junction within and without equilibrium. Explain the differences between them.

Exercise 7
Starting with the expressions for the charge carrier concentrations in terms of the Fermi level, derive Shockley equation.

Exercise 8

Explain the physical meaning of each one of the terms that appear in the definition of I_0.

Exercise 9

In Shockley equation, I_0 depends on $p_0(x_n)$ and $n_0(-x_p)$. Express these two quantities in terms of the band gap, the temperature and the concentration of dopants in each side of the junction.

Exercise 10

According to the laws of electrostatics, wherever there is an electric field, there is associated with it an electrostatic potential difference.

In a diode, there is an internal electric field, so there is an electrostatic potential difference. Suppose you have the brilliant idea of measuring this potential difference directly. So, you go to the teaching lab, get a diode straight from the drawer, and measure the potential difference with a voltmeter. To your disappointment, however, the voltmeter reading is zero. But why? Why is the voltage reading zero if there is an electrostatic potential difference across the diode?

4

Photovoltaic devices (mainly solar cells)

Learning objectives

In this chapter I will introduce the photovoltaic effect, which underlies the operation of photodetectors and solar cells, with a focus on solar cells. You will learn the basic properties of solar cells, the physical meaning of their main parameters, and the physical origins of the limitations on their efficiencies. The chapter closes with a brief discussion on connection of solar panels.

4.1 Solar cells and photodetectors

In section 3.3 we learned that illumination increases the concentration of free electrons and the concentration of holes in a semiconductor. We also learned that this simultaneous increase in the concentrations of both types of carriers splits the Fermi level into *quasi*-Fermi levels, which is the signature of a system out of thermodynamic equilibrium.

In sections 3.4 and 3.5 we studied the characteristics of a **p-n** junction wherein thermodynamic equilibrium had been broken by the application of an external potential. We learned that the external potential splits the Fermi level into *quasi*-Fermi levels, and that a diffusion current flows through the **p-n** junction when the electrostatic potential barrier is lowered by a forward bias.

If the Fermi level is split when an external voltage is applied, will an external voltage appear when the Fermi level is split due to illumination?

Essentials of Semiconductor Device Physics, First Edition. Emiliano R. Martins.
© 2022 John Wiley & Sons Ltd. Published 2022 by John Wiley & Sons Ltd.
Companion website: www.wiley.com/go/martins/essentialsofsemiconductordevicephysics

The answer to this question is a resounding and environmentally friendly yes!! The appearance of an external voltage when a diode is illuminated is an example of the photovoltaic effect, which is at the heart of crucial modern technologies such as solar cells and photodetectors. In this section, I will show the essential features of the photovoltaic effect in the context of a diode under illumination.

4.2 Physical principles

To understand the operation of an illuminated diode, let us begin by considering an open circuit condition, which means that the diode is not connected to any circuit. The energy band diagram of the diode within equilibrium (no illumination) and the diode without equilibrium (under illumination), are shown in Figure 4.1a,b. When the diode is illuminated, "extra" free electron and hole pairs are created upon absorption of photons. When these charges are generated inside the depletion region, they are swept by the internal electric field. The electric field points from **n** to **p**, so electrons are pulled towards **n**, while holes are pulled towards **p**. Thus, the generation of free electrons and hole pairs induces a drift current, whose direction is from **n** to **p**. This current is called the "photocurrent", and I designated it as I_{illum} in Figure 4.2b. Notice the direction of I_{illum}: it is the direction of the reverse biased current (from **n** to **p**).

But, hang on: aren't we considering a diode in open circuit? If the circuit is open, then there is no current flow, so how could there be a drift current? If there is a drift current, but the total current is zero, then necessarily there must be another current cancelling the drift current. Where does this other current come from? Well, the Fermi level is split, so that means that the electrochemical potential is no longer constant across the junction. We learned in the previous chapters that this is the signature of a system out of equilibrium, so we should expect a current to flow due to the splitting of the Fermi level. Indeed, the band diagram of Figure 4.1b is quite similar to the diagram of a forward biased diode. In both cases, the splitting of the Fermi level lowered the electrostatic potential. Thus, just like in a forward biased diode, the lowered barrier allows a diffusion current to flow from **p** to **n**. This diffusion current cancels the drift current caused by illumination. Consequently, no external current flows through the diode.

What about an external voltage? If we plug a voltmeter between the two sides, would we measure anything? Recall what we learned in Chapter 1: a potential difference appears between the terminals when there is a difference in the *total* potential. But the total potential is the electrochemical potential, that is, it is the Fermi level. Therefore, to assert that the Fermi level is split is

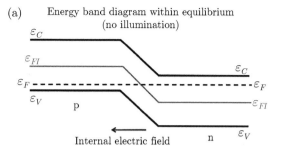

(a)   Energy band diagram within equilibrium
(no illumination)

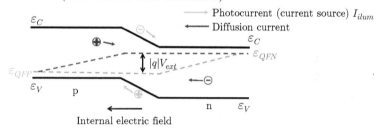

(b)   Energy band diagram without equilibrium
(semiconductor under illumination)

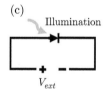

(c)

Figure 4.1 a) Energy band diagram of a p-n junction within equilibrium (no illumination). b) Energy band diagram of the p-n junction without equilibrium (under illumination). c) Open circuit representation of an illuminated p-n junction.

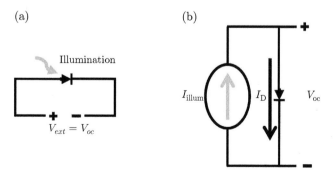

Figure 4.2 a) Open circuit representation of an illuminated diode. b) Equivalent circuit.

logically equivalent to assert that there is a potential difference between the two sides. Conclusion: the reading of the voltmeter would be V_{ext}, where V_{ext} is related to the *quasi*-Fermi levels through Equation (3.34). Notice that V_{ext} is the same external voltage that would appear if the same forward current (diffusion current) was flowing in a diode without illumination.

We have concluded that illumination creates an external voltage in a diode, which thus can act as a source, like a battery. This is an example of conversion of electromagnetic energy (the energy of the absorbed photons) into electric energy, which is the basic principle underlying the operation of photodetectors and solar cells. But to use this energy, one needs to close the circuit, for example, by connecting a load to it. When the circuit is closed, then an external current flows through it, which entails that the diffusion and drift current no longer cancel each other. This is a key feature of the operation of photovoltaic devices, and it is quite nicely illustrated by means of an equivalent circuit, which is the topic of the next section.

4.3 The equivalent circuit

The dynamics of an illuminated diode is intuitively demonstrated by means of an equivalent circuit, as shown in Figure 4.2, still considering open circuit. In the equivalent circuit (Figure 4.2), the photocurrent I_{illum} is treated as a current source, which drives the diode in forward bias. The current flowing down through the diode is the diffusion current I_D. Since the circuit is open, no current flows externally, which means that the diffusion current has the same magnitude of I_{illum} (but opposite direction, which is shown in the equivalent circuit by the fact that I_{illum} "goes up" and I_D "goes down"). The external voltage that appears in open circuit condition is called the "open circuit voltage" and is designated as V_{oc}.

Now suppose that you close the circuit by connecting a resistor (a load), as shown in Figure 4.3a. There is a voltage difference between the diode, so there will be a voltage difference across the resistor, which means that a current will flow through the resistor. This external current is designated as I_R in Figure 4.3. What is the origin of this current? Well, it can only be the photocurrent I_{illum}. Indeed, as also shown by the equivalent circuit of Figure 4.3b, part of I_{illum} drives the diode, and part of it flows through the load. So, the diffusion current I_D no longer fully cancels the photocurrent I_{illum}. That means that the diffusion current must be lower with the load than it was in open circuit. But if the diffusion current is lower, so is the external voltage, for the same reason that the current of a regular (not illuminated) diode is lower when a lower forward bias is applied as compared to a higher forward bias.

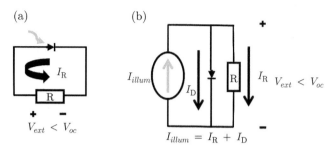

Figure 4.3 a) Circuit representation of an illuminated diode with a load. Notice the direction of current in the diode: though the voltage on the p side is higher than on the n side (as in forward bias), the current direction is from the n side to the p side (as in reverse bias). b) Equivalent circuit.

We have just learned that, when a load is connected, the diffusion current I_D must be reduced to allow an external current I_R to flow. Since I_D is the current that drives the diode (see Figure 4.3b), a lower I_D entails a lower V_{ext}. That means that, the higher the external current I_R is (and hence the lower I_D is), the lower the external voltage V_{ext} is. As the load resistance goes to zero, that is, when the diode is short-circuited, the external voltage goes to zero, and the external current is maximized. This external current is called the "short-circuit current" and is designated as I_{SC} in Figure 4.4. When the diode is short-circuited, the photocurrent flows externally, which means that $I_{SC} = I_{illum}$ and, consequently $I_D = V_{ext} = 0$.

Summarizing the two extreme conditions: in open circuit, the external current is zero, and the external voltage is maximum; in short-circuit, the external current is maximum, and the external voltage is zero.

Now, go back to Figure 4.3a. The voltage across the diode is higher on the **p** side than on the **n** side, so this voltage constitutes a forward bias. But notice that the direction of the current is from **n** to **p**, so this is a reverse current. It makes sense that this is a reverse current, since this

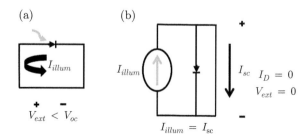

Figure 4.4 Representation of a short-circuited illuminated diode. I_{sc} is the short-circuit current.

is just the photocurrent, which, as we have just seen, is a drift current. But how could it be that the voltage of the diode is forward biased while the current is reverse biased? Well, that means that the voltage is positive and the current is negative. But if the voltage is positive and the current is negative, then the power $P = VI$ is negative! A negative power is the signature of a device that is generating power, instead of consuming. And that makes sense: the illuminated diode is acting as the source ("the battery") of the circuit. It is, in fact, converting light into voltage, which is the photovoltaic effect. That is the essence of a solar cell. And it is also the essence of a photodetector. The only difference between a solar cell and a photodetector is what you want to do with it: the former is used to generate energy, whereas the latter is used to convert light signals into electrical signals. But the underlying physical processes are the same.

4.4 The I × V curve and the fill-factor

Now that we have understood qualitatively the main features of a solar cell as captured by the equivalent circuit, let us see how we could describe it more rigorously. Take a look again at the circuit shown in Figure 4.3. We have basically three equations. One is the equation relating the current and voltage in the diode, that is, Shockley equation (Equation (3.68)):

$$I_D = I_0 \left[\exp \left(\frac{|q| V_{ext}}{k_B T} \right) - 1 \right]$$

Isolating the voltage, we get:

$$V_{ext} = \frac{k_B T}{|q|} \ln \left(\frac{I_D}{I_0} + 1 \right) \qquad (4.1)$$

That is our first equation. But we also have the good old Ohm's law, that is:

$$V_{ext} = R I_R \qquad (4.2)$$

Finally, we also have:

$$I_R = I_{illum} - I_D \qquad (4.3)$$

For a given value of I_{illum} and R, Equation (4.1), Equation (4.2) and Equation (4.3) can be solved for V_{ext} and I_R. Typically, the solution is found numerically. For example, you could express Equation (4.1) in terms of I_R by using Equation (4.3). Then you could find V_{ext} and I_R from Equation (4.1) and Equation (4.2). One way to do it numerically is to subtract Equation (4.2) from Equation (4.1) and then find the zero of this new equation. Calling the result of this subtraction $f(I_R)$, we have:

$$f(I_R) = \frac{k_B T}{|q|} \ln\left(\frac{I_{illum} - I_R}{I_0} + 1\right) - RI_R \qquad (4.4)$$

Then you can write a little computer code to find the zero of Equation (4.4), that is, the value of I_R that gives $f(I_R) = 0$. Since this value satisfy Equation (4.1), Equation (4.2) and Equation (4.3), it is the solution of the this set of equations. Once you have I_R, you can find V_{ext} from Equation (4.2). If you do that for many values of R, (ideally going from 0 to ∞), you get a plot of $I_R \times V_{ext}$. As an example, I show in Figure 4.5 the plot I obtained assuming $I_0 = 10^{-9} A$, $I_{illum} = 0.25\ A$ and room temperature:

There are some important pieces of information captured in this plot that deserve spelling out. First of all, recall that each point in this plot corresponds to a different value of R. As explained above, the current obtained when $R = 0$ is called the short-circuit current I_{SC}. So, for $R = 0$, we have $I_R = I_{SC} = I_{illum}$. At the other extreme, when $R \to \infty$, we have $V_{ext} = V_{OC}$, where V_{OC} is the open circuit voltage. But in an open circuit the current through the load is zero, that is $I_R = 0$, which entails that $I_D = I_{illum}$. Thus, from Equation (4.1), we get:

$$V_{OC} = \frac{k_B T}{|q|} \ln\left(\frac{I_{illum}}{I_0} + 1\right) \qquad (4.5)$$

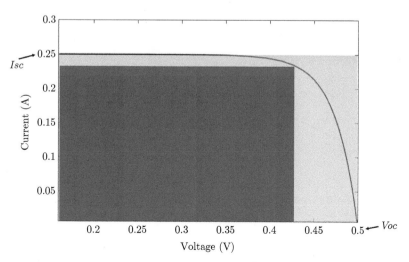

Figure 4.5 I × V curve of a solar cell assuming $I_0 = 10^{-9} A$ and $I_{illum} = 0.25\ A$.

In our example ($I_0 = 10^{-9}$ and $I_{illum} = 0.25$), Equation (4.5) gives $V_{OC} = 0.5$, as can be checked in Figure 4.5.

So, we have two extremes: one in which the current flowing through the load is maximum, but the voltage is zero, and one in which the current flowing is zero but the voltage is maximum. Since the power delivered to the load is $P = I_R V_{ext}$, both extremes result in zero power delivered. Thus, there is an optimum load that maximizes the power delivered. Such an optimum point can be seen by plotting the voltage against the power, as shown in Figure 4.6. Notice that, in our example, the power is maximized at the point $V_{ext} \approx 0.426$.

The voltage and current points of maximum power define an area in the $I_R \times V_{ext}$ plot, shown as the red square in Figure 4.5. Thus, this area is a measure of the maximum power delivered by the load. One important parameter of a solar cell is the fill-factor FF, which is defined as the ratio of the area of maximum power, to the area of the square obtained by the product $I_{SC}V_{OC}$, shown as the light blue square in Figure 4.5. Defining the current and voltages giving maximum power as I_m and V_m, respectively, the fill-factor is thus given by:

$$FF = \frac{I_m \times V_m}{I_{SC} \times V_{OC}} \qquad (4.6)$$

In our example, $I_m = 0.235\ A$, $V_m = 0.426\ A$ and $FF = 0.8$. In practice, one usually works out the maximum power $(I_m V_m)$, from FF, I_{SC} and V_{OC}.

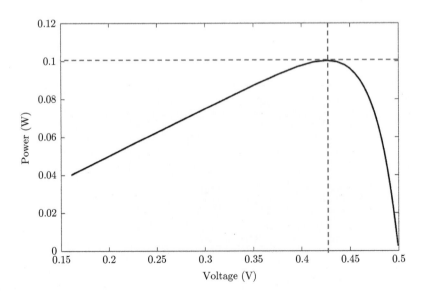

Figure 4.6 Power delivered to the load as a function of voltage.

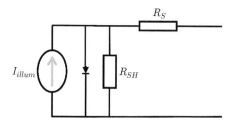

Figure 4.7 Equivalent circuit taking into account ohmic losses (through the series resistance R_S) and recombination losses (through the shunt resistance R_{sh}).

The equivalent circuit of an actual solar cell or an actual photodetector, however, is a bit more complex than what we have shown, because we have ignored losses. There are two main loss mechanisms: ohmic losses and recombination losses. Ohmic losses are due to the resistivity of the **p-n** junction and its metal contacts, and recombination losses are due to annihilation of charges due to recombination of free electron and hole pairs. In the equivalent circuit, the ohmic losses are accounted for by a series resistance R_S, as shown in Figure 4.7. Recombination losses, on the other hand, are modelled by a shunt resistance R_{SH}, since it constitutes essentially a mechanism of current leakage. An ideal solar cell has $R_S = 0$ and $R_{SH} = \infty$. The main effect of these losses is to reduce the fill-factor.

4.5   Efficiency of solar cells and the theoretical limit

The most important parameter of a solar cell is its efficiency. In the literature, the efficiency is often designated by the Greek letter η. The efficiency is defined as the ratio of the maximum power density delivered by the solar cell P_m, to the the sun's incident power density P_i:

$$\eta = \frac{P_m}{P_i} \tag{4.7}$$

The power density is the power per unit of area. According to Equation (4.6), the total power (in units of watts) is given by $FF \times I_{SC} \times V_{OC}$. To convert the total power to power density, we need to replace the total short-circuit current I_{SC} by the short-circuit density J_{SC} (recall that the current density it the current per unit of area). Thus:

$$P_m = FF \times J_{SC} \times V_{OC} \tag{4.8}$$

With the help of Equation (4.8), the efficiency can be expressed as:

$$\eta = \frac{FF \times J_{SC} \times V_{OC}}{P_i} \tag{4.9}$$

Obviously, solar incidence varies from place to place and from time to time: there are sunny days (unless you live in Scotland) and there are cloudy days. There is winter and there is summer. So, the incident solar power density P_i depends on where you are and when you are where you are. To avoid this inconvenience, solar cells are often characterized by standard power density spectra that are representative of the solar irradiance. One widely used spectrum for terrestrial applications is the Reference Air Mass 1.5 Spectra, or AM1.5G, shown in Figure 4.8. Notice that this spectrum has units of power per unit area per wavelength. So, the total power density is the integral over all wavelengths (the area of the spectrum).

As is evident from Equation (4.8) and Equation (4.9), the efficiency of a solar cell depends on the product $J_{SC} \times V_{OC}$. What can be done to maximize this product and hence the efficiency?

The answer to this question captures some important physics of solar cells. Let us begin by inspecting what J_{SC} depends on. As discussed in section 4.2, the current comes from the free charges created by absorption of photons. So, to maximize the current, we need to maximize absorption. But to be absorbed, a photon must have an energy higher than the band gap energy E_G. The energy of a photon, on the other hand, is inversely proportional to its wavelength:

$$E_f = \frac{hc}{\lambda} \tag{4.10}$$

where E_f is the photon energy, h is Planck's constant and c is the speed of light. Thus, in principle, we want a band gap sufficiently short to absorb the entire solar spectrum. For example, say that we are greedy and want to absorb the entire spectrum. By inspecting Figure 4.8, we notice that the lowest energy photons have a wavelength of ~2500 nm. So, the energy corresponding to this wavelength should set the band gap energy in this scenario. But I hasten to

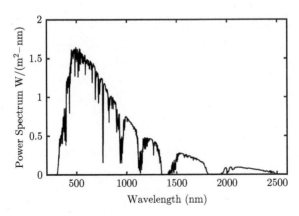

Figure 4.8 The AM1.5G solar spectrum.

warn you that this is not a good idea. Choosing a small as possible band gap does not give you the best efficiency, because the smaller the band gap is, the lower the V_{OC} is, so you may get a super-duper J_{SC} with this strategy, but you end up with a lousy V_{OC} and hence a low efficiency.

This trade-off between J_{SC} and V_{OC} is a major factor limiting the efficiency of solar cells, and it is fundamentally a consequence of the process called thermalization, which is illustrated in Figure 4.9. Suppose that a photon with energy much higher than the band gap is absorbed, creating an electron-hole pair. Due to the high photon energy, the electron lands high up in the conduction band (Figure 4.9a). But this is not a stable state, so the electron quickly relaxes to the bottom of the conduction band (Figure 4.9b). Thus, the electron loses part of the initial energy, which is transferred as heat to the underlying atomic structure (thus, the energy lost to relaxation warms up the semiconductor).

The conversion of part of the incoming photon's energy into heat is the physical process underlying the trade-of between J_{SC} and V_{OC}: to avoid thermalization, we need a high band gap, which gives us a high V_{OC}; but if the band gap is too high, we cannot absorb low-energy photons, and thus we lose J_{SC}.

You may have guessed that there must be an optimum band gap energy that balances J_{SC} and V_{OC}. Now guess who first calculated the optimum band gap. Of course, it must have been our old friend Shockley! He and

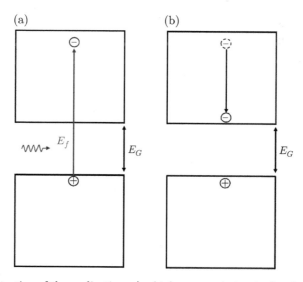

Figure 4.9 Illustration of thermalization. a) a high energy photon is absorbed, resulting in a high energy electron in the conduction band. b) the electron relaxes to the bottom of the conduction band, losing energy as heat.

his pal Hans J. Queisser published a classic paper in 1961 where they not only calculated the optimum band gap, but also worked out the maximum possible efficiency of a solar cell. This is the so-called Shockley-Queisser limit. They found that the optimum band gap is 1.1. $e.\ V.$ and the maximum efficiency is about 30%. Now, silicon has a band gap of about $1.12 - 1.14.\ e.\ V.$; so not too bad!

Maybe you have heard of super expensive solar cells reaching 75% efficiencies. These cells do exist, and the reason why they can go beyond the 30% efficiency limit is that they consist of a stack of solar cells (the technical term for a stack of solar cells is a "multijunction"). By stacking solar cells, you can make each of them absorb just a part of the spectrum, thus reducing the thermalization problem and allowing to overcome the theoretical limit of 30%.

4.6   Connections of solar cells

To conclude this chapter, suppose that a certain solar cell panel delivers a voltage $1\,V$. But suppose you want twice as much voltage, that is, $2\,V$. So, you decide to buy two panels and to set them up in series, as shown in Figure 4.10a. You measure the total voltage and, unsurprisingly, you have your $2\,V$. What is the problem of this connection? Suppose that something happens to one of the diodes. It does not need to be anything dramatical, it could be just that a big bird recovering from the curry the other night pooped on one of the panels, thus blocking the illumination. Now, you have one illuminated diode in series with a non-illuminated smelly diode. Without the illumination, the pooped diode becomes a regular diode. But the current flow is in the reverse direction, so the regular diode is reverse biased. We learned in Chapter 3 that a reverse biased diode is essentially an open circuit. That means that the photocurrent of the clean diode can no longer flow through the circuit. So, just because accidentally one of the diodes got in the dark, the entire system stopped working. That is why connections in series tend to be problematic: if one device stops working, the entire system stops working.

Figure 4.10 Solar panels connected in series. a) When both panels are illuminated, the total voltage delivered by the system is the sum of each individual voltage. b) If the illumination of one panel is compromised, the entire system stops working, because the non-illuminated panel becomes a regular diode operating in reverse bias, thus essentially opening the circuit.

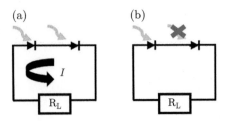

4.7   Suggestions for further reading

Here I presented the essential concepts of photovoltaics. If you want to learn more, I recommend as a starting point the always instructive website *pveducation.org*. If you want to go deeper, then I recommend Peter Wurfel's: *Physics of Solar Cells: from principles to new concepts.*

4.8   Exercises

Exercise 1
Explain the physical origin of the photovoltaic effect.

Exercise 2
Use Shockley equation to find an expression for the open circuit voltage of a solar cell in terms of the photocurrent.

Exercise 3
Sketch a graph showing the current (vertical axis) against the voltage (horizontal axis) of a solar cell. Identify the short-circuit current and open circuit voltage in your graph.

Exercise 4
Do you expect the optimum band gap of a solar cell to be dependent on the solar spectrum? Why?

Exercise 5
Consider the I x V curve of an ideal solar cell. Now imagine that a series resistance is attached to it. Which part of the I x V curve will be mostly affected by the presence of the series resistance? Hint: to check if your answer is correct, you can use the I x V curves simulators available at pveducation.org.

Exercise 6
Find an equation for the current of a solar cell considering a shunt resistance, but neglecting the series resistance. How do you expect that the shunt resistance affects the I x V curve?

Exercise 7
Suppose you need to measure the series and shunt resistances of a solar cell. Suppose you choose to do the measurement in the dark, that is, without illumination. In this case, the solar cell becomes a diode in parallel with the shunt resistance and in series with the series resistance. Do you expect the series resistance to be more important when the diode is forward biased or reverse biased? What about the shunt resistance?

5

Transistors

Learning objectives

In this chapter, you will be introduced to an important electronic component: the transistor. The basic physical processes are discussed, alongside their electrical properties and two paradigmatic applications: transistors as amplifiers and transistors as electronic switches, with an example of application in logic gates.

As a last example of application of **p-n** junctions, I will describe the essential principles of transistors. Transistors are used to make amplifiers and are also the building blocks of logic gates, which are the building blocks of microprocessors. Thus, it is hard to overemphasize the importance of transistors in modern society.

A transistor is a device with three terminals. The idea is to use one terminal to control the current flowing between the other two terminals. The two most important types of transistors are the Bipolar Junction Transistor (BJT) and the Metal Oxide Semiconductor Field Effect Transistor (MOSFET). For pedagogical reasons, I begin by showing the physical principles of the BJTs and how they can be used as amplifiers. Then, I discuss the main properties of MOSFETs, and give an example of their application in logic gates.

Essentials of Semiconductor Device Physics, First Edition. Emiliano R. Martins.
© 2022 John Wiley & Sons Ltd. Published 2022 by John Wiley & Sons Ltd.
Companion website: www.wiley.com/go/martins/essentialsofsemiconductordevicephysics

5.1   The Bipolar Junction Transistor (BJT)

There are two types of BJTs: **n-p-n** and **p-n-p**. If you understand the operation of one type, you will also understand the operation of the other. It is more common to describe the **n-p-n**, so that is also my choice.

5.1.1   Physical principles of the BJT

The structure of an **n-p-n** BJT, alongside the symbol used in circuit drawing, are shown in Figure 5.1a. The transistor is formed by two **p-n** junctions. There are three terminals: the emitter, the collector and the base. The base terminal is connected to the **p** doped region. Usually, the **p** doped region is much narrower than the others, and it is only lightly doped. The doping of the **n** regions, however, is much heavier than in the **p** region. Consequently, transport in an **n-p-n** BJT is dominated by free electrons.

The transistor can operate in basically three modes: the cut-off mode, the active mode and the saturation mode. In the cut-off mode, all junctions are reverse biased (or not sufficiently forward biased to conduct). In the active mode, the Base-Emitter (BE) junction is forward biased (see Figure 5.1), whereas the Base-Collector (BC) junction is reverse biased (see Figure 5.1). Finally, in the saturation mode, both junctions are forward biased.

The active mode is the operation mode of interest, so we focus attention on it. Figure 5.1b shows the polarization scheme of the **n-p-n** in active mode: the BE junction is forward biased, while the BC junction is reverse biased. The forward bias polarization induces diffusion of electrons from the emitter to the base. These electrons diffuse all the way to the second depletion region, between the base and the collector. When these electrons reach the depletion region of the reverse biased base-collector junction, the internal electric field "collects" these electrons, that is, it pulls them towards the collector. Thus, the free electrons are emitted by the emitter and are collected by the collector. This process is represented in Figure 5.2.

As mentioned earlier, the main purpose of a transistor is to control the current between two terminals through the third. Specifically, we want to control the current that flows from the collector to the emitter by means of the voltage V_{BE}, applied between the base and the emitter. Thus, we need to find an expression for this current as a function of V_{BE}. Since the electrons emitted by the emitter are collected by the collector, this current is essentially the current of the emitter-base junction (there is a correction though, which I will discuss in the next section). As explained in Chapter 3, this is a

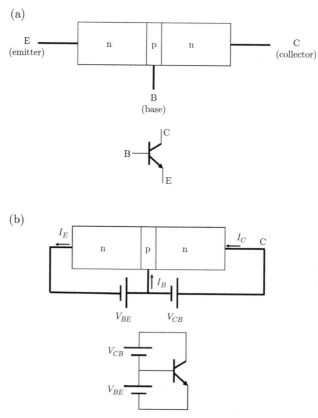

Figure 5.1 a) Structure of a Bipolar Junction Transistor of the n-p-n type. The inset shows the symbol used in circuit drawing. b) BJT in active mode.

diffusion current, so, to find it, we first need to obtain the concentration of charge carriers.

To find the concentration, consider the axis of Figure 5.3. The width of the region between the two depletion regions is designated as W. The origin of the coordinate axis is placed at the edge between the **p** region and the first depletion region. The free electrons reaching this region diffuse until the second depletion region. We already know that transport by diffusion in the presence of recombination results in an exponential decay of charge carriers (Equation (1.110) or Equation (1.111)). Due to the light doping in the **p** region, the recombination is very low, and, as you will prove in the exercise list, in the limit of low recombination the exponential becomes a line. So, we can approximate the concentration spatial dependence to a line. Importantly, since the electric field collects the charges arriving at the second

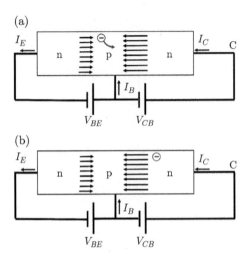

Figure 5.2 Representation of charge transport in a BJT in active mode. The arrows represent the direction of the internal electric fields in the two depletion regions. a) The emitter-base junction is forward biased, allowing diffusion of electrons from the emitter (n doped) to the base (p doped). The electrons diffuse across the base until they reach the second depletion region, between the base and collector. b) the electric field in the second depletion region collects the electrons.

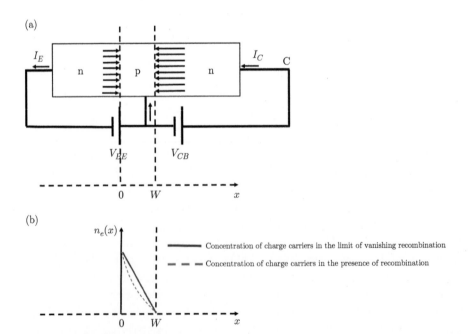

Figure 5.3 Concentration of minority charge carriers (electrons) in the base (p region). a) Geometry delimitating the p region. b) charge concentration profile in the p region.

depletion region, it imposes the boundary condition that the charge concentration is zero at the point $x = W$.

We have thus determined that the charge carrier concentration is a line, and that it satisfies the boundary condition $n_e(W) = 0$. Therefore, its spatial dependence within the **p** region is given by:

$$n_e(x) = -\frac{n_e(0)}{W} x + n_e(0) \tag{5.1}$$

On the other hand, the diffusion current is:

$$J = |q| D_N \frac{dn_e}{dx} \tag{5.2}$$

Therefore:

$$J = -\frac{|q| D_N}{W} n_e(0) \tag{5.3}$$

Notice that the current is negative (electrons move in the direction of positive x, so the charge current is in the direction of negative x).

We still need to express $n_e(0)$ in terms of the voltage in the junction. Since $n_e(0)$ is the concentration of excess minority carriers, its dependence on the voltage across the junction is given by Equation (3.63). Adapting to the axis of Figure 5.3, we get:

$$n_e(0) = n_0(0) \left[\exp\left(\frac{|q| V_{ext}}{k_B T}\right) - 1 \right] \tag{5.4}$$

where $n_0(0)$ is the equilibrium concentration of electrons in the **p** region.

Notice that, according to Figure 5.2 and Figure 5.3, $V_{ext} = V_{BE}$. Therefore:

$$n_e(0) = n_0(0) \left[\exp\left(\frac{|q| V_{BE}}{k_B T}\right) - 1 \right] \tag{5.5}$$

Finally, substituting Equation (5.5) into Equation (5.3), we find:

$$J = -\frac{|q| D_N n_0(0)}{W} \left[\exp\left(\frac{|q| V_{BE}}{k_B T}\right) - 1 \right] \tag{5.6}$$

Equation (5.6) expresses the essence of a transistor: the current between two terminals (emitter and collector) is controlled by the voltage in the third terminal V_{BE}. Typically, the exponential factor is much larger than one, so the current can be approximated by:

$$J = -\frac{|q|D_N n_0(0)}{W}\left[\exp\left(\frac{|q|V_{BE}}{k_B T}\right)\right] \tag{5.7}$$

This is, of course, the current density. If we want to know the total current, all we need to do is to multiply it by the cross-sectional area A. Thus:

$$I = -\frac{A|q|D_N n_0(0)}{W}\left[\exp\left(\frac{|q|V_{BE}}{k_B T}\right)\right] \tag{5.8}$$

In the same spirit of Chapter 3, we encapsulate all the parameters multiplying the exponential by defining the saturation current I_0:

$$I_0 = \frac{A|q|D_N n_0(0)}{W} \tag{5.9}$$

Finally, notice that the negative sign in Equation (5.8) arises because our axis points to the right side, but the emitter and collector currents flow to the left. As we are interested in the magnitude of the current, we drop the negative sign and keep in mind that the current enters the transistor through the collector and exists through the emitter. With these conventions, the current can be conveniently expressed as:

$$I = I_0\left[\exp\left(\frac{|q|V_{BE}}{k_B T}\right)\right] \tag{5.10}$$

5.1.2   The β parameter and the relationship between emitter, collector and base currents

Ideally, Equation (5.10) describes both emitter and collector currents, and the base current is zero. However, in reality, there is a small base current. This current is just the hole current of the forward biased BE junction (recall that this current is much smaller than the emitter current because the base is only lightly doped and very narrow). The important point to notice is that, since the base current is just the current of holes in a forward biased **p-n** junction, it must also be proportional to $\exp\left(\frac{|q|V_{BE}}{k_B T}\right)$. Thus, it must also have the form of Equation (5.10), but with a different (and much smaller) saturation current. Thus, since all currents are proportional to $\exp\left(\frac{|q|V_{BE}}{k_B T}\right)$, they must all be proportional to each other.

The proportionality between the currents defines an important parameter in a BJT transistor: the β parameter. It is defined as the proportion between the collector current I_C and the base current I_B:

$$I_C = \beta I_B \qquad (5.11)$$

Since $I_C \gg I_B$, β must be a large number (typically $\beta \approx 100$). This is one of the parameters that engineers check in the datasheet of transistors.

We can also relate I_C to the emitter current I_E via the β parameter. Recall that the collector and base currents enter the transistor, while the emitter current exits it (see Figure 5.2). Since there can be no accumulation of charges in the transistor, we must have:

$$I_E = I_B + I_C \qquad (5.12)$$

With the help of Equation (5.11):

$$I_E = \frac{I_C}{\beta} + I_C = \frac{\beta + 1}{\beta} I_C$$

Notice that, if $\beta = 100$, then $I_E = 1.01 \times I_C$, that is, $I_E \approx I_C$, as expected.

Box 14   The p-n-p BJT transistor

The structure of a **p-n-p** BJT is shown in Figure 14.1a, together with its symbol. Notice that it is the same structure as the **n-p-n**, but with the roles of **p** and **n** doping reversed. Thus, the physics of **p-n-p** BJTs are essentially the same as in the **n-p-n**, but instead of transport dominated by electrons (as in the **n-p-n**), transport in the **p-n-p** is dominated by holes.

As in the **n-p-n**, there are three modes of operation: cut-off, active mode and saturation. As in the **n-p-n**, in the cut-off mode both junctions are cut-off (reverse biased), in the active mode the BE junction is forward biased and the BC is reversed biased, and in saturation all junctions are conducting (forward biased).

The **p-n-p** BJT in active mode is shown in Figure B14.1b. Notice that the BE junction is forward biased and the BC is reverse biased, as expected.

(Continued)

(Continued)

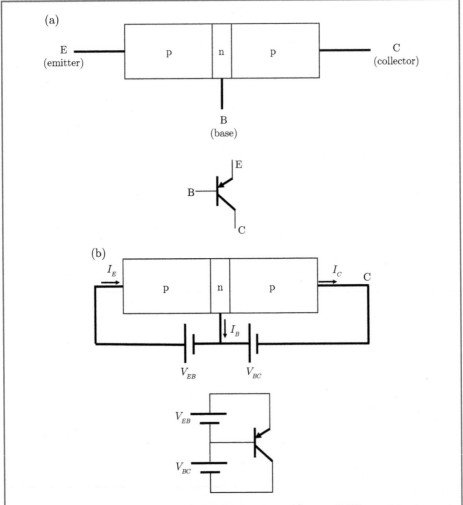

Figure B14.1 a) structure of the p-n-p BJT transistor. b) p-n-p BJT transistor in active mode.

5.1.3 Relationship between I_C and V_{CE} and the Early effect

A key feature of BJT transistors is the relationship between the collector current I_C and the voltage between collector and emitter V_{CE}. This relationship is illustrated in Figure 5.4a. To understand the main features of this plot, take a look at the circuit of Figure 5.4b, which represents an experiment that could be made to obtain the plot of Figure 5.4a. In this experiment, V_{BE} is kept fixed, and the collector current is measured for various values of V_{CE}, thus obtaining the plot of Figure 5.4a. To avoid unnecessary notation

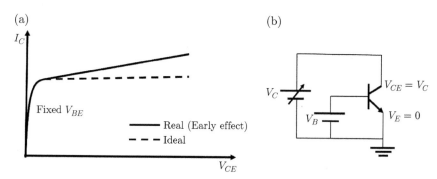

(a)

I_C

Fixed V_{BE}

——— Real (Early effect)

– – – Ideal

V_{CE}

(b)

V_C

V_B

$V_{CE} = V_C$

$V_E = 0$

Figure 5.4 a) Illustration of the relationship between the collector current and the voltage between collector and emitter. b) circuit used to explain the relation between I_C and V_{CE}.

cluttering, I assume that the emitter is grounded, so that $V_E = 0$ and, consequently, $V_{CE} = V_C$, and $V_{BE} = V_B$. Notice that the BE junction is forward biased with a fixed voltage V_B.

Assuming fixed V_B, imagine that we begin with $V_C = 0$. If $V_B > 0$ and $V_C = 0$, then $V_B > V_C$, which means that the BC junction is also forward biased. Thus, for small V_C, the diode is in saturation mode (I will say a few words about this regime soon). When V_C has been increased sufficiently to reverse bias the BC junction, the transistor enters the active mode, and I_C obeys Equation (5.10). According to Equation (5.10), the current depends only on V_{BE}. Thus, since we are keeping V_B fixed, the current should remain constant as V_C is increased. This is the ideal case, as shown in Figure 5.4a (dashed line). There is, however, a small slope (which I exaggerated in the plot to emphasize it, see Figure 5.4a – solid line). The physical origin of this slope is captured by the W term in Equation (5.9). Recall from Figure 5.3 that W is the width of the **p** region between the two depletion regions. As V_C increases, the BC junction gets more and more reversely biased, which means that its depletion region widens, thus squeezing the **p** region, that is, reducing W. But the smaller W is, the higher the saturation current I_0 is. Therefore, the increase in I_C with V_C is due to the dependence of I_0 on V_C through the width W. This is called the Early effect.

As promised, a few words about the saturation regime: you may be wondering why the current in the saturation regime (small V_{CE} – Figure 5.4a) is smaller than the current in the active mode. The reason is very simple: suppose we are in the active mode and start decreasing V_C towards the saturation regime. Recall that, in the active mode, the BC junction is reverse biased, and in the saturation mode it is forward biased. So, what happens to the current when the saturation regime is reached? Of course, the BC junction starts to conduct, so there is a new current in the transistor. But this new current, that is, the current due to the forward biased BC junction, flows from the base to the collector, so it exits the transistor through the collector. In other words, the new current is in the opposite direction of I_C. Therefore, the new current subtracts from the "original", active mode I_C, and the result is a reduction in I_C.

5.1.4 The BJT as an amplifier

An important application of BJT transistors is in amplifiers. A commercial amplifier is typically a complicated circuit with lots of transistors, but the essential idea can be conveyed with a very simple circuit, as shown in Figure 5.5a. The circuit has a source V_0 connected to a resistor R, which is in turn connected to the transistor collector. The output voltage V_{out} is taken between the collector and the emitter. We are interested in checking how V_{out} depends on V_{BE}.

Let us first take a qualitative look at the dependence of V_{out} on V_{BE}. In the region of small V_{BE}, V_{out} is constant and equal to V_0, as shown in Figure 5.5b. This is because the transistor is in the cut-off mode, that is, all junctions are reverse biased (V_{BE} is too small to forward bias the BE junction), so no current flows through the collector. As V_{BE} becomes sufficiently high to forward bias the BE junction, the transistor enters the active mode, and a collector current I_C flows through the transistor. This current causes a voltage drop across the resistor, thus reducing V_{out}. If we keep increasing V_{BE}, there will be a point where the BC junction also gets forward biased, and the transistor enters the saturation mode. We are interested in the active mode region, which lies between these two regimes, and is marked by dashed vertical lines in Figure 5.5b.

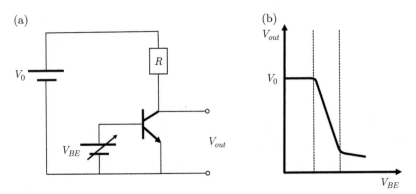

Figure 5.5 a) amplifier circuit. b) Output voltage against V_{BE}. The dashed vertical lines delimitate the active mode region.

Applying Kirchhoff's voltage law to the circuit of Figure 5.5a gives:

$$V_0 = RI_C + V_{out} \qquad (5.13)$$

Where I_C is the collector current. In the active mode, the collector current obeys Equation (5.10). Applying this equation to Equation (5.13) and solving for V_{out}, we get:

$$V_{out} = V_0 - RI_C = V_0 - R\left[I_0 \exp\left(\frac{|q| V_{BE}}{k_B T}\right)\right] \qquad (5.14)$$

Thus, the circuit of Figure 5.5a turned the exponential dependence between current and voltage in a **p-n** junction into an exponential dependence between V_{out} and V_{BE}. Therefore, a narrow variation in V_{BE} causes a wide variation in V_{out}. It is this feature that is explored in amplifiers.

The usual goal of an amplifier is to amplify an alternating signal (for example, your favourite song). In Figure 5.6a, v_{in} is the signal to be amplified. Notice that it is superimposed on a constant (DC) component, which I called V_{BEQ}. The Q comes from "quiescent point". This term designates the operation point, that is, the point in the V_{out} vs V_{BE} curve around which the voltage oscillates when the signal is applied. An example of Q point is marked by a dot in the curve of Figure 5.6b, where a graphical representation of amplification is shown. The sinusoidal oscillation in the horizontal axis represents the signal to be amplified (the input signal v_{in}), and the sinusoidal oscillation in the vertical axis represent the signal amplified, which I designated as v_{out} in Figure 5.6a.

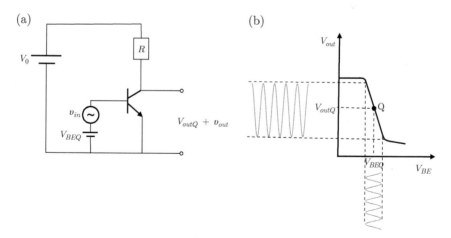

Figure 5.6 Amplification using BJTs. a) amplifier circuit. The signal to be amplified is applied to the BE junction, and the output signal is taken between the collector and the emitter. b) Representation of amplification of a single frequency signal.

Notice that, to avoid distortion, the Q point must be chosen as in the middle of the active region as possible. Indeed, if v_{in} is too high, the operation point may approach the edges of the active region (or even cross them), which leads to distortion in v_{out}, and your favourite song being spoiled. Since this is a key concept in amplifiers, let us take a closer look at the relationship between amplification and distortion.

First of all, what is distortion? When you amplify your favourite song to play it as loud as possible to dance to it as if there was no tomorrow, you definitely want the amplified song to be the same as the original one, only louder. Mathematically, that means that you want v_{out} to be proportional to v_{in}. Anything that deviates from a proportionality factor is bad news. But, if v_{out} is to be proportional to v_{in}, then the curve inside the active region in Figure 5.6b must be a line. Alas, it is not a line, it is an exponential (Equation (5.14)), so distortion is unavoidable. Let us quantify it against the amplification gain. To do it, go back to Equation (5.14). With the identity $V_{BE} = V_{BEQ} + v_{in}$ (see Figure 5.6), Equation (5.14) reads:

$$V_{out} = V_0 - RI_0 \exp\left[\frac{|q|\left(V_{BEQ} + v_{in}\right)}{k_B T}\right]$$

Now we isolate the DC component from the input signal:

$$V_{out} = V_0 - RI_0 \exp\left(\frac{|q| V_{BEQ}}{k_B T}\right) \exp\left(\frac{|q| v_{in}}{k_B T}\right)$$

and express the output signal V_{out} explicitly in terms of the DC component V_{outQ} and the output signal v_{out} ($V_{out} = V_{outQ} + v_{out}$):

$$V_{outQ} + v_{out} = V_0 - RI_0 \exp\left(\frac{|q| V_{BEQ}}{k_B T}\right) \exp\left(\frac{|q| v_{in}}{k_B T}\right) \qquad (5.15)$$

The DC component V_{outQ} is the output at the Q point, that is, when $v_{in} = 0$. Thus, according to Equation (5.15):

$$V_{outQ} = V_0 - RI_0 \exp\left(\frac{|q| V_{BEQ}}{k_B T}\right) \qquad (5.16)$$

Now we need an expression for v_{out}. We are particularly interested in separating the signal from the distortion. In other words, we want to separate the contribution that is proportional to v_{in} from the rest. A Taylor expansion around $v_{in} = 0$ is just right for this job:

$$\exp\left(\frac{|q| v_{in}}{k_B T}\right) \approx 1 + \frac{|q|}{k_B T} v_{in} + \frac{1}{2}\left(\frac{|q|}{k_B T}\right)^2 v_{in}^2 + higher\ order\ terms \quad (5.17)$$

For our purposes, it is sufficient to truncate the expansion in the second order term (recall that v_{in} must be small to keep the oscillation within the active region, and the smaller v_{in} is the more negligible the higher order terms are). Thus, substituting Equation (5.17) into Equation (5.15):

$$V_{outQ} + v_{out} \approx V_0 - R I_0 \exp\left(\frac{|q| V_{BEQ}}{k_B T}\right)\left[1 + \frac{|q|}{k_B T} v_{in} + \frac{1}{2}\left(\frac{|q|}{k_B T}\right)^2 v_{in}^2\right]$$

With the help of Equation (5.16), we conclude that:

$$v_{out} \approx - R I_0 \exp\left(\frac{|q| V_{BEQ}}{k_B T}\right)\frac{|q|}{k_B T} v_{in} - R I_0 \exp\left(\frac{|q| V_{BEQ}}{k_B T}\right)\frac{1}{2}\left[\frac{|q|}{k_B T}\right]^2 v_{in}^2$$

$$(5.18)$$

The first term on the right-hand side of Equation (5.18) is proportional to v_{in}, so this is the output signal of interest. The other term is not proportional to v_{in}, but to v_{in}^2, so it is an unwanted second order contribution, that is, it is the distortion (or noise). Since the distortion depends on v_{in}^2, the smaller v_{in} is, the less significant its contribution is.

The proportionality between v_{out} and v_{in} defines the gain G of the amplifier:

$$G = \frac{v_{out}}{v_{in}} \quad (5.19)$$

According to Equation (5.18), the gain of the amplifier of Figure 5.6 is:

$$G = - R I_0 \exp\left(\frac{|q| V_{BEQ}}{k_B T}\right)\frac{|q|}{k_B T} \quad (5.20)$$

The negative sign expresses the fact that this is an inverting amplifier, that is, the input and output signal are out of phase by 180°. This is evident from Figure 5.6b: V_{out} decreases when V_{BE} increases (and vice-versa).

Notice that there is not much room for the engineer to improve the amplifier using the simple circuit of Figure 5.6a. For example, if the engineer wants to increase the gain, she or he could increase either R or V_{BEQ}, but that would also increase the distortion in the same proportion (see Equation (5.18)). Thus, it is the goal of the circuit designer to complicate the circuit aiming at optimizing the gain and minimizing the distortion.

5.2 The MOSFET

The Metal Oxide Semiconductor Field Effect Transistor is the typical transistor of choice to make the logic gates used in microprocessors. In logic gates, which is arguably its most important application, the transistor acts as an electronic switch. In this chapter, I will show you the basic ideas underlying the physical operation of the MOSFET, with a focus on its application as an electronic switch and in logic gates. The treatment is simplified and qualitative. More thorough treatments are suggested in section 5.3.

5.2.1 Physical principles

Like BJTs, MOSFETs also come in two types: the **n**-channel MOSFET, also called NMOS, and the **p**-channel MOSFET, also called PMOS.

The structure of an NMOS is shown in Figure 5.7a: it consists of a **p** type body (the substrate), with two heavily **n** doped regions (the heavy doping is represented by the superscript + in n^+). The transistor has four terminals, but usually the body terminal and the source terminal are short-circuited (see Figure 5.7a), so we consider only three terminals: the gate G, the source S and the drain D. The source and drain are connected to the highly doped **n** regions, whereas the gate is isolated from the transistor by a thin (of the order of a few nanometres) insulating layer. Typically, the insulator is made of a dioxide, most commonly silicon dioxide; hence the term Metal Oxide Semiconductor.

The idea of the MOSFET is to control the conductivity between S and D by means of G. Notice that, due to the insulator, no current enters or exists through G. With no voltage applied, the resistivity between S and D is very high, since there are two back-to-back **p-n** junctions, formed by the body and each heavily doped **n** region. These two back-to-back **p-n** junctions can be understood as two back-to-back diodes connected between S and D, as represented on the right-hand side of Figure 5.7a. Thus, with low or no voltage on the gate, the transistor can be treated as an open switch (high resistance between S and D). The symbol for the NMOS used in circuits is shown on the right-hand side of Figure 5.7a.

When a sufficiently high voltage is applied to the gate, as represented in Figure 5.7b, the electric field associated with the gate voltage expels holes

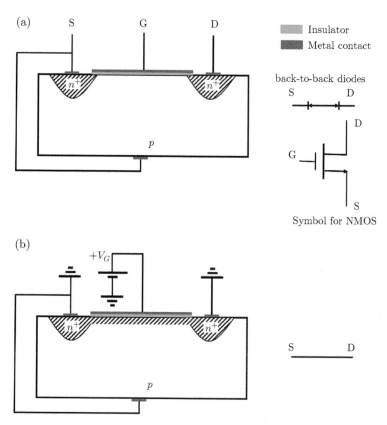

Figure 5.7 a) Physical structure of an NMOS, with the body and source short-circuited. The transistor consists of two back-to-back p-n junctions, as represented on the right-hand side. b) When a positive bias is applied between the gate and the other terminals, the two p-n junctions are simultaneously forward biased, thus making the resistivity between source and drain very low, which allows treating them approximately as short-circuited, as represented on the right-hand side. The low resistance is due to the creation of a channel with high concentration of free electrons, that is, an n-channel.

away from the region under the insulator, and it attracts free electrons from the heavily doped **n** regions to the region under the insulator. Thus, a channel of free electrons, or **n**-channel, is formed between the drain and the source. The high concentration of free electrons guarantees a low resistivity between S and D; thus, in this condition, the S and D terminals can be treated approximately as short-circuited, as represented on the right-hand side of Figure 5.7b. In terms of the equivalent two diode model, the positive voltage on the gate forward biased both back-to-back diodes at the same time. The low resistivity between S and D means that the NMOS acts as a closed switch (a short-circuit) when a sufficiently high voltage is applied to the gate. Thus, the NMOS is an electronic switch: it is an open switch (high resistivity) for

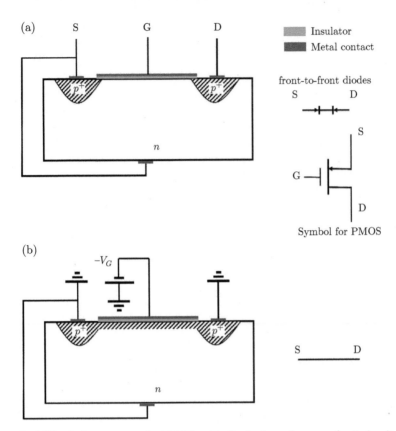

Figure 5.8 a) Physical structure of a PMOS, with the body and source short-circuited. The transistor consists of two front-to-front p-n junctions, as represented in the inset. b) When a negative bias is applied between the gate and the other terminals, the two p-n junctions are simultaneously forward biased, thus making the resistivity between source and drain very low, which allows treating them approximately as short-circuited, as represented in the inset. The low resistance is due to the creation of a channel with high concentration of holes, that is, a p-channel.

low voltage on the gate, and it is a closed switch (low resistivity) for high voltage on the gate.

The structure of the PMOS transistor is similar to the NMOS, but now the doping profile is inverted. As shown in Figure 5.8a, the body is **n** doped and the regions under source and drain are heavily **p** doped. Thus, the PMOS can be understood as forming two front-to-front diodes. Again, these two diodes guarantee a very high resistance between S and D when there is no voltage difference between G and the other terminals.

If a bias is applied to the gate so that its voltage is sufficiently lower than the voltage at S and D, as shown in Figure 5.8b, both diodes will be simultaneously forward biased, and the terminals S and D can be treated approximately as short-circuited. In this situation, the electric field of the negatively biased gate expels free electrons from the body and attracts holes from the heavily doped regions, thus forming a channel with high concentration of holes, or a **p**-channel. The high concentration of holes in the channel guarantees a low resistance between S and D.

Thus, the PMOS also acts as an electronic switch, but now the switch is closed when the voltage at the gate is sufficiently lower than the voltage at S and D (as opposed to the requirement of higher voltage on the gate for NMOS transistors).

One useful way to remember how to close the switch is to look at the arrows in the symbols for NMOS and PMOS (Figure 5.7 and Figure 5.8). In both cases, the switch is *open* when the voltage on the *tip* of the arrow is higher than the voltage on the *tail*. Likewise, in both cases, the switch is closed when the voltage on the *tail* of the arrow is higher than the voltage on the *tip*. In the next section, we look at two paradigmatic applications of MOSFET: logic inverters and logic gates.

5.2.2 Examples of applications of MOSFETs: logic inverters and logic gates

A logic inverter is a device with two terminals: one input and one output. As implied by the name, the inverter turns a high signal into a low signal, and vice-versa. In the language of logic, that means turning 1 into 0, and vice-versa, or True into False, and vice-versa.

The structure of a logic inverter using MOSFETs is shown in Figure 5.9a. It consists of a source (V_{HIGH} in Figure 5.9a) connected to the source of a PMOS, whose drain is connected to the drain of an NMOS, whose source is connected to a low voltage V_{LOW}, represented as the ground in Figure 5.9a. The input is connected to the gates of both transistors, and the output is connected to the drains. Now we analyse what happens when a high voltage is applied to the input (Figure 5.9b) and then when a low voltage is applied to the input (Figure 5.9c).

If a high voltage is applied to the input (Figure 5.9b, left part), then the PMOS opens, but the NMOS closes (recall the rule: for the transistor to close, the voltage at the tail of the arrow must be higher than the voltage at the tip of the arrow). Thus, the opened PMOS isolates the source from the output, whereas the closed NMOS connects the ground to the output. The equivalent circuit, treating the PMOS as an open circuit and the NMOS as a short-circuit, is shown on the right-hand side of Figure 5.9b. To conclude: if a high voltage is applied to the input, a low voltage appears at the output, as expected.

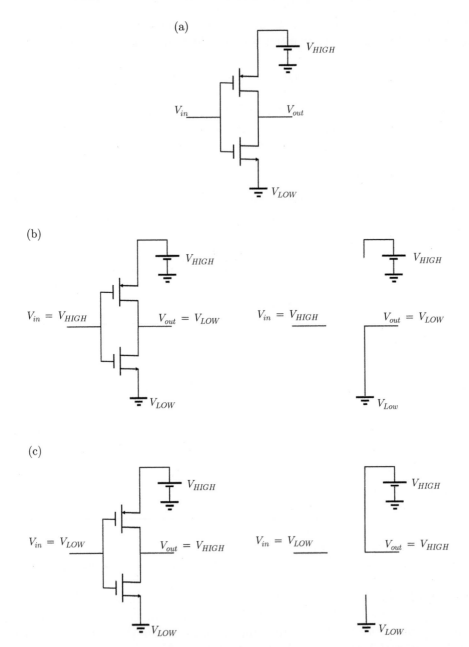

Figure 5.9 a) logic inverter based on MOSFETs. b) left side: when a high voltage is applied to the input, the PMOS opens and the NMOS closes; consequently, a low voltage appears in the output; right side: equivalent circuit. c) left side: when a low voltage is applied to the input, the PMOS closes and the NMOS opens; consequently, a high voltage appears in the output; right side: equivalent circuit.

If, on the other hand, a low voltage is applied to the input (Figure 5.9c, left part), then the PMOS closes and the NMOS opens, as represented by the equivalent circuit on the right-hand side of Figure 5.9c. Thus, the closed PMOS connects the source to the output, whereas the open NMOS isolates the ground from the output. Consequently, a high voltage appears at the output when a low voltage is applied to the input, as expected.

Logic inverters are often used in logic gates. As an example of a logic gate using a logic inverter, consider the circuit of Figure 5.10. The circuit highlighted by the dashed box is an inverter. The input of the inverter, labelled $\overline{C}$, is connected to a circuit with four transistors, labelled T1, T2, T3 and T4. Overall, there are two inputs, labelled A and B, and one output, labelled C.

The circuit of Figure 5.10 is an AND logic gate, whose truth table is shown as an inset on the rightmost side of Figure 5.10. Notice that T1 and T2 are PMOS transistors, while T3 and T4 are NMOS transistors. Input A is connected to T1 and T3, while input B is connected to T2 and T4.

Let us check if this circuit indeed performs an AND operation. To simplify, I will denote high voltage by H and low voltage by L.

According to the arrangement of the circuit of Figure 5.10, when A is H, T1 opens and T3 closes. When B is H, T2 opens and T4 closes. When A is L, T1 closes and T3 opens. When B is L, T2 closes and T4 opens.

On the one hand, since T3 and T4 are connected in series, the only way that $\overline{C}$ can be connected to the ground is if both T3 and T4 are closed. Thus,

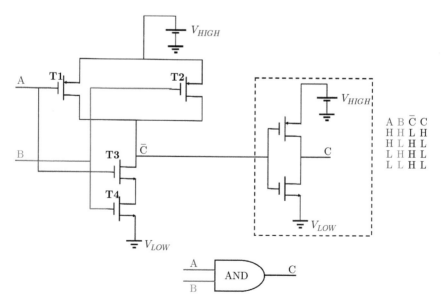

Figure 5.10 A logic gate performing the AND operation.

the only way $\overline{C}$ can be L is when both A and B are H. Furthermore, when both A and B are H, then both T1 and T2 are open, which means that the source is isolated from $\overline{C}$.

We have concluded that, when both A and B are H, then $\overline{C}$ is L. But $\overline{C}$ is the input of the inverter, whereas C is the output, which means that C is H when $\overline{C}$ is L. Thus, we conclude that, if both A and B are H, then C is H.

On the other hand, since T1 and T2 are connected in parallel, the source connects to $\overline{C}$ when either of them, or both of them, are closed. Thus, if either A or B are L, then $\overline{C}$ is H. Likewise, if both A and B are L, then $\overline{C}$ is also H. Furthermore, if either A or B are L, then either T3 or T4 is open, thus isolating $\overline{C}$ from the ground (of course, $\overline{C}$ is also isolated from the ground when both A and B are L).

We have concluded that, if either A or B are L, or if both are L, then $\overline{C}$ is H and, consequently C is L.

Summarizing: the output is H only if both A and B are H, which is the characteristic of an AND operation. See the truth table on the right-hand side of Figure 5.10.

5.3 Suggestions for further reading

If your main interest is to delve further into the physics of transistors, then I suggest *Physics of Semicondutor Devices*, by S. M. Sze. If, however, your main interest is in electronic circuits, then I suggest *Microelectronic circuits*, by Sedra/Smith.

5.4 Exercises

Exercise 1
Prove that, in the limit of zero recombination (infinite lifetime), the spatial variation of the concentration of charge carriers when transport is by diffusion is linear.

Exercise 2
At room temperature ($T = 300$ K), the collector current of a BJT is $1mA$ when the voltage between base and emitter is 0.7V. Find the saturation current I_0.

Exercise 3
Consider the I x V curve of Figure 5.4a. In the active region, the collector current I_C depends linearly on V_{CE}, and thus can be approximated as:

$$I_C = AV_{CE} + B$$

Find a relationship between the width W (see Figure 5.3) and V_{CE} in terms of the parameters A and B

Exercise 4
Consider again the I x V curve of Figure 5.4a, with the linear approximation

$$I_C = AV_{CE} + B$$

Make a sketch of the curve of Figure 5.4a, together with the linear approximation, but allow the line defined by the equation above to be extended beyond the active region. Find the point where this line crosses the horizontal axis and express this point in terms of A and B. This point is an important parameter of BJTs, and it is called the Early voltage. Remarkably, it does not depend on V_{BE}, so that if you have a second line, for another V_{BE}, it will also cross the horizontal axis at the same point.

Exercise 5
Typically, the BE junction of a BJT conducts when $V_B - V_E > 0.5$ V, and the BC junction conducts when $V_B - V_C > 0.4$ V. Close to saturation, the voltage drop in the BE junction is about ~ 0.7 V, which entails that the saturation point is reached when $V_C - V_E \sim 0.7 - 0.4 = 0.3$ V. Assume $I_0 = 10^{-15} A$.

Now, consider the amplifying circuit of Figure 5.5, with $V_0 = 15$ V and $R = 10$ kΩ.

(a) Find the values of V_{BE} at the edges of the active region (see Figure 5.5b).

(b) Find the collector current, and output voltage at the cut-off point, at the saturation point and at the Q point chosen in the middle of the horizontal axis ($V_{BE} = (0.7242 + 0.5)/2 = 0.612$ V) (see Figure 5.6b).

(c) Now repeat the design, but choosing the Q point to lie in the middle of the vertical axis, that is, $V_{CEQ} = 7.5$ V.

Exercise 6

(a) Consider the equation for the gain of the BJT amplifier. Show that the gain has no units.

(b) Calculate the gain assuming the conditions of Exercise 5c.

Exercise 7

Consider the amplifier of Figure 5.6. Suppose that, within a certain temperature range, all parameters are fixed (this is just an idealization). In this case, do you expect the signal to noise ratio to improve or worsen with the temperature?

Exercise 8

Which logic operation is performed by the circuit below?

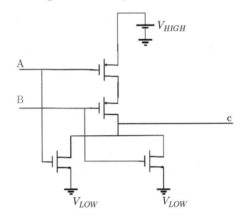

Appendix: Geometrical interpretation of the chemical potential and free energy

According to Equation 1.40, which I rewrite below for convenience, the variation of energy in a system can be expressed as a sum of two terms:

$$\Delta U = T\Delta S + \mu \Delta N$$

The first term $T\Delta S$ is the heat contribution, and the term $\mu \Delta N$ is the free energy. In the literature, it is often asserted that heat is the part of the energy that carries entropy, while the free energy is the part of the energy that is free of entropy. In this section, I offer a simple and intuitive geometrical interpretation of these contributions and assertions.

As discussed in Chapter 1, the entropy S is a function of energy U and number of particles N. By expressing S as function of U and N, we are treating U and N as the free variables. But we can equivalently consider U as a function of S and N, thus treating S and N as the free variables. We shall adopt this latter point of view here.

Suppose we have curve describing U as a function of S for fixed N, as shown in Figure A1.1. We can envisage a second curve of U against S, but now for fixed $N + \Delta N$. If ΔN is a small quantity, then the two curves are very close to each other, which means that they are virtually identical (of course, we are ignoring the possibility of discontinuities). Thus, considering small variations allows us to treat both curves as identical, but with a vertical offset between them due to the contribution of ΔN to the total energy, as shown in Figure A1.1. Notice that the vertical offset has been exaggerated to facilitate visualization of the geometrical parameters.

We again assume that the system is in thermal, but not diffusive, contact with a reservoir. Moreover, we assume that the initial energy is U_0, as located by the red dot in Figure A1.1.

Essentials of Semiconductor Device Physics, First Edition. Emiliano R. Martins.
© 2022 John Wiley & Sons Ltd. Published 2022 by John Wiley & Sons Ltd.
Companion website: www.wiley.com/go/martins/essentialsofsemiconductordevicephysics

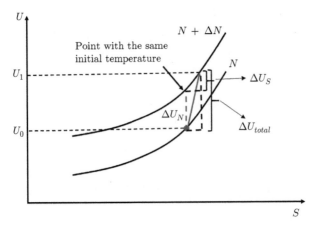

Figure A1.1 Representation of the energy as a function of entropy and number of particles, highlighting two curves of energy U against entropy S for two values of fixed N, separated by a small quantity ΔN.

Now, consider that the system received ΔN new particles (imagine that you quickly opened the system, quickly threw in ΔN particles, and quickly closed it again), and that these ΔN new particles carried a total energy ΔU_{total}. Thus, these new particles make the system jump to the point in the upper curve where the energy is $U_1 = U_0 + \Delta U_{total}$, as indicated by the tip of the red arrow in Figure A1.1.

We can break down ΔU_{total} into two contributions, as also shown in Figure A1.1: a contribution due solely to the change in the number of particles ΔU_N (as indicated by the dashed blue vertical line), and a contribution due to the change in entropy ΔU_S. Thus:

$$\Delta U_{total} = \Delta U_S + \Delta U_N$$

Notice that these two contributions are identical to the ones obtained from calculus, that is:

$$\Delta U_{total} = \frac{\partial U}{\partial S}\Delta S + \frac{\partial U}{\partial N}\Delta N$$

where

$$\Delta U_S = \frac{\partial U}{\partial S}\Delta S$$

and

$$\Delta U_N = \frac{\partial U}{\partial N}\Delta N$$

Thus, Figure A1.1 is only a geometrical representation of these two well-known terms from calculus of two variables.

If the process were isolated from the universe, then the energy of the system would remain at U_1 forever. But, since we are considering a system in thermal contact with a reservoir, the system will lose part of this energy until it reaches the point with the same temperature as the initial point (which is also the temperature of the reservoir).

Recall that the temperature is the derivative of U against S, which means that the temperature is the tangent of the curve. As the two curves are virtually identical, just with a vertical offset between them, then the point at the upper curve that has the same tangent as the initial point will be the point with a vertical offset from the initial, as indicated in Figure A1.1. Therefore, though the system jumped to the tip of the arrow when it received the ΔN particles, it will relax back to the point in the upper curve that has the same tangent of the initial point, thus losing part of its energy. Since this final point is vertically offset, the final energy difference is ΔU_N. In other words, the energy that is left after thermal equilibrium is reached is ΔU_N; thus ΔU_N is the free energy. Furthermore, the system had to lose the part of the energy "that carried entropy", that is, the part ΔU_S, which was due to a change in the entropy. This latter part is the heat.

Now we need to show that these terms coincide with the definitions given in Chapter 1 for heat and free energy. The heat part is obvious: since

$$\Delta U_S = \frac{\partial U}{\partial S} \Delta S$$

and $\dfrac{\partial U}{\partial S} = T$, it follows immediately that:

$$\Delta U_S = T \Delta S$$

Thus, we confirm that $T \Delta S$ is indeed the part of the energy that carries entropy and that is lost to the reservoir when the system goes back to thermal equilibrium – in short: it is the heat.

Now we need to prove that ΔU_N is indeed the free energy.

To prove this, consider again the same curves as before, but now focusing attention on the right triangle, with angle θ, adjacent side $-\Delta S$, and opposite side ΔU_N, as shown in Figure A1.2 (the negative sign in $-\Delta S$ is due to the fact that ΔS is negative in the horizontal transition from the initial point to the curve for $N + \Delta N$, but the side of a triangle has to be a positive quantity).

Thus, considering the triangle, we have:

$$\tan \theta = \frac{\Delta U_N}{-\Delta S}$$

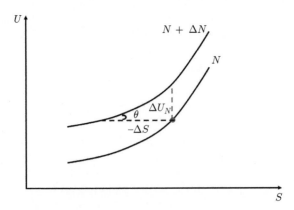

Figure A1.2 Geometrical representation of the chemical potential.

but

$$\tan \theta = \frac{\partial U}{\partial S} = T$$

therefore:

$$T = \frac{\Delta U_N}{-\Delta S}$$

On the other hand:

$$\Delta U_N = \frac{\partial U}{\partial N} \Delta N$$

therefore:

$$T = \frac{\frac{\partial U}{\partial N} \Delta N}{-\Delta S}$$

Rearranging, we find that:

$$\frac{\partial U}{\partial N} = -T \frac{\Delta S}{\Delta N}$$

Notice that ΔS is a horizontal variation, that is, with U fixed. Therefore, for infinitesimal variations, the relation above becomes:

$$\frac{\partial U}{\partial N} = -T \frac{\partial S}{\partial N}$$

Thus ΔU_N can be expressed as:

$$\Delta U_N = \frac{\partial U}{\partial N}\Delta N = -T\frac{\partial S}{\partial N}\Delta N$$

In Chapter 1, we defined the potential as:

$$\mu = -T\frac{\partial S}{\partial N}$$

Therefore, we conclude that:

$$\Delta U_N = \mu \Delta N$$

This concludes the proof that the free energy, as defined in chapter 1, indeed coincides with ΔU_N.

One last comment: in the literature, the free energy is often expressed in terms of the "Helmholtz free energy" F, which is defined as:

$$F = U - TS$$

From this definition, it follows that:

$$\Delta F = \Delta U - \Delta TS - T\Delta S = \mu\Delta N - S\Delta T$$

where

$$\frac{\partial F}{\partial N} = \mu$$

and

$$\frac{\partial F}{\partial T} = -S$$

Thus, F is a function of N and T.

In a process held at constant temperature, $\Delta T = 0$, thus

$$\Delta F = \mu\Delta N$$

So, the free energy can be understood as the energy that is available in a thermal process held at constant temperature.

Index

absolute zero, 63, 64

absorption, 6, 81, 140, 172, 180

acceptors, 117

active mode, 186–188, 191–195

AM1.5G, 180

amplification, 195, 196

amplifier, 185, 194–197, 205, 206

AND operation, 203, 204

anisotropic materials, 74

area. *see* cross-sectional area

atmosphere, 29, 83

atom(s). *see* doping; impurity

atomic orbitals, 88, 123. *see also* orbitals

atomic structure, 70, 94, 98, 102, 112, 114, 167, 181

average, 45, 48, 49, 82

band, energy, 92, 115, 118, 128, 129, 145, 148, 149, 168, 172, 173. *see also* conduction band; valence band

band gap, 94, 98–112, 114, 115, 121, 122, 140, 166, 167, 169, 180–183. *see also* direct band gap; indirect band gap

band theory, 87

barrier, 143, 148, 171, 172

base, 186–188, 190–192, 194

base-collector, 186. *see also* BC junction

base current, 190, 191

base-emitter, 186. *see also* BE junction

BC junction, 186, 193, 194, 205. *see also* base-collector

BE junction, 186, 190, 191, 193–195, 205

bipolar junction transistor. *see* BJT

BJT, 185–188, 191, 192, 194, 195, 198, 204, 205. *see also* n-p-n BJT; p-n-p BJT

body, 198–201

Boltzmann constant, 20, 114

boron, 117

boundary condition(s), 5, 80, 137, 189

box, 3–5, 18, 21, 26, 33, 42, 45–50, 54, 56, 62, 65–67, 69–76, 78, 83, 94, 98, 99, 104–106, 113, 118, 121, 123, 144–146, 162, 166, 191, 203

built-in potential, 130, 133, 135, 143

capacitor, 39–41, 128, 130, 133, 147, 148, 162

cavity, 3–5, 7

centre of mass, 90

channel, 198–201. *see also* n-channel; p-channel

charge(s), 1, 38–42, 67, 68, 76, 80, 84, 87, 94–96, 117–120, 126–132, 135–138, 143, 144, 148, 150–152, 155, 157, 158, 168, 172, 179, 180, 187–189, 191, 204

charge carrier(s). *see* free charge carriers

charge, elementary, 40, 42

charge, net, 38, 84, 95, 117

charges, concentration, 1, 76, 131, 144. *see also* density, charges

chemical potential, 30–42, 62, 67, 69, 74, 83–85, 102, 123, 126, 128–130, 143, 144, 207, 210. *see also* electrochemical potential

chemistry, 88

classical mechanics, 6

classical system, 2

coins, 8–12, 15–18, 25–27, 29, 31, 49, 65, 66

collector, 186–195, 204, 205

colour, 81, 167

conduction band, 94–100, 102, 104, 110–112, 115–118, 123, 124, 130, 135, 140, 150, 155, 166, 181

conductivity, 94, 96, 145, 153, 168, 198

Essentials of Semiconductor Device Physics, First Edition. Emiliano R. Martins.
© 2022 John Wiley & Sons Ltd. Published 2022 by John Wiley & Sons Ltd.
Companion website: www.wiley.com/go/martins/essentialsofsemiconductordevicephysics